Dynamic Structure of NREM Sleep

Péter Halász • Róbert Bódizs

Dynamic Structure of NREM Sleep

Péter Halász, M.D., Ph.D., DSCi
Institute of Experimental Medicine and
Institute of Neuroscience
Budapest
Hungary

Róbert Bódizs, Ph.D.
Semmelweis University
Institute of Behavioral Science
Budapest
Hungary

ISBN 978-1-4471-4332-1 ISBN 978-1-4471-4333-8 (eBook)
DOI 10.1007/978-1-4471-4333-8
Springer London Heidelberg New York Dordrecht

Library of Congress Control Number: 2012949899

Printed on acid-free paper

Springer is part of Springer Science+Business Media (www.springer.com)

Foreword

Peter Halász and Robert Bódizs has taken the lead in studies of non REM sleep regulation. They continues the noble tradition of Giuseppe Moruzzi and Horace Magoun, Dominic Purpura and Mircea Steriade and seeks, as they did, the neurophysiological rules governing the brain in its control of states of consciousness. For a variety of very sound reasons their work focuses on the initial, so-called non REM, phase of sleep. What is new and truly exciting about their work is their discovery that there are wheels with wheels in the generation even of this apparently monotonic state.

We used to think of sleep as the absence of waking. Now, thanks to Halász and Bódizs, we are able to appreciate the enormous complexity of what we hoped was a simple, if active process. One specific feature is the cyclic alternating pattern by which the brain changes its activity within the non REM phase. This mechanism, called CAP, was discovered in Parma Italy by Terzano and his group. CAP has received little attention outside of Parma and it is important that Halász and Bódizs have corrected that neglect. In doing so they has gone a long way to elucidating the mechanisms of this fascinating dynamic.

It is probably true that most persons in the world know little or nothing about the third of their life that they spend asleep. Even neurobiologists have tended to ignore sleep. But the study of sleep may be a promising avenue to understanding our supreme gift, consciousness. We now also know that sleep is fraught with peril and its understanding is essential to the maturation of sleep disorders medicine.

My pleasure is to recommend this book as the very latest and most promising scientific assaults upon our most important organ, the brain in one of least well understood states, sleep.

Allan Hobson

Acknowledgements

We would like to thank the work of all our colleagues co-operating with us in those earlier periods when the work on the sleep microstructure have begun. Their contribution is present not only by their papers reviewed here but obviously in the form of their view incorporated in the book. We would like to mention János Ujszászi, who was the first authors' main co-worker in the K-complex research, Péter Rajna, who co-authored several of our papers and contributed with original ideas in the development of the dynamic microstructural view, Ivan Pal, who led our hands in those early heroic periods when we started to learn digital processing.

We are very grateful for the 27 years long co-operation and friendship to Parma Sleep Research Group and personally to Mario Giovanni Terzano and Liborio Parrino in elaborating mutually fruitful concepts of microstructural sleep dynamism and multiple aspect of relationship between sleep and epilepsy. We learned a lot from them.

A special thank is due to Alexander Borbély and Lopes da Silva who red the manuscript and provided critical remarks and encouredgement for us.

We thank to all sleep researchers who fertilized our thinking by their papers and ideas. If we claim things they would not agree, the eventually improper interpretation is exclusively our responsivity.

Special thanks belong to our colleague Anna Szűcs, who made a great job by improving the English text, correcting our unclear sentences with competent professionalism and provided proper wording to our intentions.

Ruben Rial Planas provided us irreplaceable help with his figures about different aspects of the reptile sleep/wake EEG.

The first author received emotional and mental support from his wife Julia Gádoros child psychiatrist, who devoted herself to time-taking discussions helping to clear up vague issues.

Contents

Abbreviations

'A'	Ascending slope (of sleep cycles)
ADA	Adenosine deaminase
AMPA	2-amino-3-(5-metil-3-oxo-1,2-oxazol-4-il)propánsav
ASDA	American Sleep Disorders Association
ATP	Adenosine triphosphate
BDNF	Brain-derived neurotrophic factor
BOLD	Blood
CAmKII	Ca^{2+}/calmodulin-dependent protein kinases II
CAP	Cyclic alternating pattern
CES	Cortical electrical stimulation
CSD	Current source density
'D'	Descending slope (of sleep cycles)
DC	Direct current
DMN	Default mode network
DSP-4	*N*-(−2-chloroethyl)-*N*-ethyl-2-bromobenzylamine
ECoG	Electrocorticography
EMG	Electromyography
ERP	Evoked response
FFT	Fast Fourier transformation
fMRI	Functional magnetic resonance image
GABA-ergic	Gamma-aminobutyric acidergic
GluR1	Glutamine receptor 1
IL1-β	Interleukin-1 beta
LTD	Long-term depression
LTP	Long-term potentiation
MnPN	Median preoptic nuclei
MUA	Multiplex unit activity
NF-kB	Nuclear factor kappa-light-chain-enhancer of activated B cells
NREM	Non-REM (sleep)
P2R	Purine receptor

PAT	Phases d'activation transitoire
PET	Positron emission tomography
RAS	Reticular arousal system
rCBF	Regional cerebral blood flow
REM	Rapid eye movement (sleep)
SPECT	Single-photon emission computer tomography
SWA	Slow-wave activity
TMS	Transcranial magnetic stimulation
TNF-α	Tumor necrosis factor alpha
VLPO	Ventrolateral preoptic area

Introduction

Waking and sleep appertain to the same part of an animal inasmuch as they are opposites.
(Aristotle)

New Trends in Sleep Research

The actuality of the recent work is that it is the very period of research when long ago described classical graphoelements of sleep EEG started to have an underlying cellular and network basis. Furthermore, we start to know the relationship among these hitherto isolated graphoelements providing possibility to put together the complex system and make assumptions for the biological role of it.

An important change in ideas about sleep is reflected in views whether sleep is a "top-down" or "bottom-up" process. After the discovery of the circadian pacemaker and hypnogenic brain systems, sleep was tend to held as a centrally regulated active process. Nowadays, the pendulum is swinging again to the opposite direction: sleep is more and more held to be a bottom-up process, determined by use of dependent metabolic and plastic demands, as this will be detailed later.

Leading to further changes in our concept, two basic assumptions have been challenged in the last years about sleep. One is connected with the discrete (homogenous) nature of sleep (and wake) states. Especially in some of the brain diseases, elements of different vigilance states can occur simultaneously. For example in, "REM behaviour disorder" (associated often with Parkinson disease) motor actions occur during REM sleep. The recognition of these "mixed states" led to the term "status dissociatus" (Mahowald and Schenk 1991).

The other basic assumption challenged is the global nature of sleep. This has been challenged by several aspects. One is the experience of local sleep differentiating the state of the different cortical columns (Krueger and Obál 1993). Another approach is showing local changes of (delta) sleep intensity after local utilization of body functions during the previous day (Kattler et al. 1994). Alternating sleep of the two hemispheres discovered in dolphins and other aquatic mammals (Mukhametov et al. 1976; Lyamin et al. 2008) pushed thinking also toward the possibility of local/regional sleep in the brain.

The higher temporal and spatial resolution by which we can study brain waves have led also to the recognition of "dissociated" phenomena in a smaller time scale. The description of "up and down states" of bellow 1-Hz oscillations in deep sleep (Steriade et al. 1993) has shown that even within a 1-Hz window, the functional states may alternate on the subsecond scale between almost wake-like to extreme deactivated states.

The message of the contemporary sleep research we try to translate for the readers is that using the words of Tononi and Cirelli (2003) "sleep is the price we have to pay for plasticity," to see sleep as an amazing tool for brain recuperation, or as Shakespeare wrote in Macbeth,

Sleep that knits up the ravell'd sleave of care,
The death of each day's life, sore labour's bath,
Balm of hurt minds, great nature's second course,
Chief nourisher in life's feast.

Macbeth (2.2.46-51)

Some Introductory Notes from the Point of View of "Systems Theory"

According to the biological application of systems theory, organism of living beings works as an open system in dynamic connection with their environment (Bertalanffy 1940). One of the elementary conditions of survival is to maintain continuous connection with the outer world (reactibility) but in the same time, to a certain sense, separate themselves from the influence of the outer world preserving steady state (homeostasis).

Sleep is an unstable (bilabile) open system during which, although the communication with the outer world is limited, the essential difference between comatose state, when arousability is partially or totally lost, is that in sleep, arousability is preserved and the possibility to change the sleep state by external sensorial input is maintained.

Chronobiology teaches that the animal organism is characterized by cyclic changes. During these cyclic changes, the organism arrives at different "states." The connection with and separation from the surrounding world are an important feature of states

Hess was the first who conceptualized the biological organisms according to this principle (Gloor 1954). He recognized the animal existence as states when activity level changes: in certain periods, it is high, ready to expend energy to fight, to escape, or express predatory action, while in other periods, activity level is low, giving place for restoration of functions. High activity was named by Hess as "ergotropic" and low as "trophotropic-endophylactic" states. Ergotropic states have been characterized by sympathetic and trophotropic-endophylactic as parasympathetic dominance. Wakefulness and sleep are characteristic examples of ergotropic versus trophotropic states.

In the same time, these seemingly opposite states are not possible to define without each other, so they are essentially – as, for example, light and darkness – complementary (interconnected and interdependent) principles. In the ancient Chinese philosophy, "yin-yang" are the symbol of such kind of twin principles with opposite features. What is more, yin-yang constantly interact, never existing in absolute stasis. For instance, dropping a stone in a calm pool of water will simultaneously raise waves and lower troughs between them, and this alternation of high and low points in the water will radiate outward until the movement dissipates and the pool is calm once more. Yin-yang, thus, are always opposite and equal qualities. Further, whenever one quality reaches its peak, it will naturally begin to transform into the opposite quality.

Since the work of Aserinsky and Kleitman (1953), we know that sleep constituted by cycles in which the relative rest periods of NREM sleep alternate with REM periods characterized by more activity. Taking a closer look, the level of vigilance also oscillates during wakefulness. Therefore, both sleep and wakefulness can be described as different sets of states, and we have learned as a principle with utmost importance that macrostates are build up by smaller microstates.

The state principle became one of the most important concepts of neurobiology. The concept was borrowed originally from the systems theory and is analogous with the states of machines. States (of existence) may characterized by a set of variables signalizing the functional state of an organism, which remains constant during a certain time period and returns in the same form from time to time. The behavioral pattern of a state is underlain by a functional state of certain systems of the organism and determines the mode of connection with the environment. The state determines the input and output characteristics of the system. For example, the neuronal networks respond to inputs differently, depending on their state (Nadim et al. 2008). Later, we will see how the state dependency determines the answers to external stimuli during different states of the sleep/wake system along the slopes of the sleep cycles. The variables of a state make probable the existence of a constellation in which a set of parameters behave predictable similarly. Behavior of an organism is possible to describe as continuous transitions of states underlain by a functional state of different interconnected neural networks determining them.

The state concept, as mentioned above, has been very useful in the growing understanding of parasomnias, especially in REM behavior disorder (Schenk et al. 1986), and promoted the application of the "state dissociation" concept in the mechanism of parasomnias and also in explaining symptoms of narcolepsy. It has an application in general when in any states there is a mixture of parameters belonging classically to two (or more) states, as an explanatory principle in different pathological mental states (e.g., multiple personality).

The state principle promoted to recognize the so-called default states of the brain. Besides the classical three kinds of existence (wake/NREM/REM), now, we differentiate between attentional, goal-directed behavior, and a peculiar behavioral state when individuals are not focused on external environment, instead they are engaged in internally focused tasks including autobiographical memory retrieval and envisioning the future. The long-ago existing suspicion of such working mode of our brain is now firmly supported by fMRI data (Buckner et al. 2008) underlain by an interconnected network of anatomical structures, having relevance in the understanding of mental disorders (autism, schizophrenia, Alzheimer disease).

Sleep, or at least NREM sleep, should be conceived as a special default state of the brain: the activity is "endophylactic" serving "trophotropic" purposes. It is interesting to consider a certain line of evolution from the stimulus response like working mode of primitive organisms, where the stimulus without any special work-up gets an immediate response toward different degrees of work-up after the arrival of the stimulus, until the development of cognitive-type work-up in human beings. The default state obviously is serving on one side the elaboration in the mental (cognitive) faculty and, on the other, the separation of the individual merging into a state supporting mental elaboration.

Sleep and Oscillations (Rhythms)

Since the discovery of the human electroencephalogram by Berger, we know that oscillation is a universal phenomenon in the brain. There are several oscillators, and certain frequency-specific oscillations are characteristic to certain neuronal assemblies having nowadays more or less known functional significance. With the advent of the digital EEG not hampered by the physical inertia of the paper-pen registration, a late revolution in the understanding of EEG rhythms went through, discovering a hitherto unknown fast (ripple) activity until 600 Hz and an infraslow activity bellow 0.1 Hz. The fast activity has evolved importance in cognitive functions and epilepsy and the infraslow activity in large-scale organization of brain activity in wakefulness and sleep.

Among the wake rhythms, the alpha rhythm is the most studied but still not the most understood, underlain by an unperturbed state of the first-order thalamic nuclei and primary sensorimotor cortical fields. A similar "off-duty" oscillation is the so-called mu rhythm over the somatosensory cortices, desynchronized by clenching of the hand, recently associated with the mirror neuron function (Perry et al. 2011). The hippocampal theta rhythm has a long research story and functionally connected with spatiotemporal orientation, an essential element of episodic memory (Cornwell et al. 2012). Hippocampal theta rhythm also connected to cognitive functions in REM sleep, and possibly similar rhythmic slow activity (delta instead of theta) is present in human beings as well (Bódizs et al. 2001). During sleep, other oscillations take place. Positive occipital sharp transient rhythms over the occipital regions after falling asleep, sleep spindles in two types (frontocentral slower and centroparietal faster variants) as the product of the burst-firing working mode of the thalamocortical system mainly in stage 2 NREM, and delta activity (detailed later) during stage 3–4 NREM in more hyperpolarized condition of the thalamic relay neurons. Further, on the top of all these, a <1-Hz slow oscillation envelops the 1–4-Hz delta and higher range oscillations. Last but not least, the importance of an infraslow oscillation of about 0.01–0.1 Hz was recognized recently. This latter is not specific to sleep but seems to organize all the state-dependent faster oscillations and shape the periodic fluctuation in neuronal excitability, providing a potential electrophysiological signature of the default mode network (Monto et al. 2008; Raichle et al. 2007). The functions of these (sleep) rhythms are slowly more and more unfolded and will be treated in more detail later in this book.

Beyond the elementary sleep EEG rhythms, sleep is characterized by other rhythms embedding sleep in the circadian rhythmicity and showing a coalition of several rhythms. They are the ultradian cycle, where NREM and REM sleep alternate; the cyclicity of the NREM stages in recurring sequences from evening to morning; and the smaller-scale oscillations unanimously detectable by macro-EEG evaluation in the form of cyclic alternating pattern (CAP), described in the mideighties by the Parma Group (Terzano et al. 1985). There are hints that this latter form of EEG alternation during sleep is in fact related to the previously mentioned infraslow oscillation (Lőrincz et al. 2009).

So sleep is characterized by interdigitated coalition of different grade oscillations, and we can say that larger oscillations are built up by smaller oscillations during the whole course of sleep. This oscillatory nature of sleep is very much accentuated in sleep contrasting to the wake state. In wake state, level of vigilance fluctuations are also present, but the variability and the amplitude of the changes are far less so conspicuous. Sleep and even NREM sleep alone are richer in texture compared to wakefulness, and this provides a good terrain for comparison in the researched dynamical relations.

Besides the coalescence of oscillations, it is important to emphasize the relationship between the amplitude and the frequency of the oscillations. The negative correlation between the frequency and amplitude of the oscillations means that slow rhythms are of larger amplitude than fast ones. The decreasing linear trend of the power spectra of the EEG in double logarithmic plots is a clear indication of this feature (Feinberg et al. 1984). The power spectra can be described by a $1/f^{\alpha}$-type function where f means frequency, while α is an exponent between 1 and 4 determining the decline of the spectra, thus the relationship between slow and fast components in the EEG (Freeman and Zhai 2009). Higher exponents are characteristic features of deeper sleep and were hypothesized to reflect the intensity of the recuperative functions of sleep (Feinberg et al. 1984). Potential reasons for this power law relationship between EEG amplitude and frequency are phenomena like self organized criticality, fractal nature of brain oscillations, or filtering properties of the extracellular media (Buzsáki and Draguhn 2004). Whatever the routes of the power law functions are, it is clear that under physiological conditions, the EEG amplitude in general is mainly determined by low-frequency components (slow rhythms). Moreover, a steeper decline in the spectra is a measure of EEG synchronization. The coalescence of the oscillations is tightly related to this property of the neural activity, as slow-frequency components can embed faster ones only if they are of higher amplitude.

References

Aserinsky E, Kleitman N. Regularly occurring periods of eye motility, and concomitant phenomena, during sleep. Science. 1953;118(3062):273–4.

Bertalanffy Lv. Der organismus als physikalisches system betrachtet. Die Naturwissenschaften. 1940;28:521–31.

Bódizs R, Kántor S, Szabó G, Szűcs A, Erőss L, Halász P. Rhythmic hippocampal slow oscillation characterizes REM sleep in humans. Hippocampus. 2001;11:747–53.

Buckner RL, Andrews-Hanna JR, Schacter DL. The brain's default network anatomy, function, and relevance to disease. Ann N Y Acad Sci. 2008;1124:1–38.

Buzsáki G, Draguhn A. Neuronal oscillations in cortical networks. Science. 2004; 304(5679):1926–9.

Cornwell BR, Arkin N, Overstreet C, Carver FW, Grillon C. Distinct contributions of human hippocampal theta to spatial cognition and anxiety. Hippocampus. 2012. doi: 10.1002/hipo.22019.

Feinberg I, March JD, Fein G, Aminoff MJ. Log amplitude is a linear function of log frequency in NREM sleep eeg of young and elderly normal subjects. Electroencephalogr Clin Neurophysiol. 1984;58(2):158–60.

Freeman WJ, Zhai J. Simulated power spectral density (PSD) of background electrocorticogram (ECoG). Cogn Neurodyn. 2009;3(1):97–103.

Gloor P. Autonomic functions of the diencephalon; a summary of the experimental work of Prof. W. R. Hess. AMA Arch Neurol Psychiatry. 1954;71(6):773–90.

Kattler H, Dijk DJ, Borbély AA. Effect of unilateral somatosensory stimulation prior to sleep on the sleep EEG in humans. J Sleep Res. 1994;3(3):159–64.

Krueger JM, Obál F. A neuronal group theory of sleep function. J Sleep Res. 1993;2(2):63–9.

Lörincz ML, Geall F, Bao Y, Crunelli V, Hughes SW. ATP-dependent infra-slow (<0.1 Hz) oscillations in thalamic networks. PLoS One. 2009;4(2):e4447.

Lyamin OI, Manger PR, Ridgway SH, Mukhametov LM, Siegel JM. Cetacean sleep: an unusual form of mammalian sleep. Neurosci Biobehav Rev. 2008; 32(8):1451–84.

Mahowald MW, Schenck CH. Status dissociatus – a perspective on states of being. Sleep. 1991;14(1):69–79.

Monto S, Palva S, Voipio J, Palva JM. Very slow EEG fluctuations predict the dynamics of stimulus detection and oscillation amplitudes in humans. J Neurosci. 2008;28(33):8268–72.

Mukhametov LM, Supin AIa, Strokova IG. Interhemispheric asymmetry of cerebral functional states during sleep in dolphins. Dokl Akad Nauk SSSR. 1976;229(3): 767–70.

Nadim F, Brezina V, Destexhe A, Linster C. State dependence of network output: modeling and experiments. J Neurosci. 2008;28(46):11806–13.

Perry A, Stein L, Bentin S. Motor and attentional mechanisms involved in social interaction–evidence from mu and alpha EEG suppression. Neuroimage. 2011; 58(3):895–904.

Raichle ME, Snyder AZ. A default mode of brain function: a brief history of an evolving idea. Neuroimage. 2007;37(4):1083–90.

Schenck CH, Bundlie SR, Ettinger MG, Mahowald MW. Chronic behavioral disorders of human REM sleep: a new category of parasomnia. Sleep. 1986;9(2): 293–308.

Steriade M, Nunez A, Amzica F. A novel slow (<1 Hz) oscillation of neocortical neurons in vivo: depolarizing and hyperpolarizing components. J Neurosci. 1993; 13:3252–65.

Terzano MG, Mancia D, Salati MR, Costani G, Decembrino A, Parrino L. The cyclic alternating pattern as a physiologic component of normal NREM sleep. Sleep. 1985;8(2):137–45.

Tononi G, Cirelli C. Sleep and synaptic homeostasis: a hypothesis. Brain Res Bull. 2003;62(2):143–50.

Chapter 1
Development of the Concept of Sleep-Wake-Promoting Systems in the Brainstem and Hypothalamus

Abstract A main part of fundamental knowledge about sleep regulation roots in the recognition of reciprocally antagonistic sleep-wake-promoting influences originating in brainstem/hypothalamic neuronal networks and in their dynamic interplay. All the presently known concepts and models of sleep regulation are essentially based on the anatomy and function of these subcortical networks. From the beginning of this century, the Harvard School of sleep research and the Lyon sleep research group have introduced important concepts about the working of the flip-flop model of "sleep switch" governing sleep/wake alternations and tried to connect these views with the earlier ideas explaining sleep cyclicity. These results have suggested that the subcortical neuronal assemblies have indeed a governing role in the EEG and behavioral changes during the sleep/wake cycle. However, within the NREM sleep period, robust dynamical changes take place, about which the flip-flop model has nothing to say about. NREM sleep cannot be taken as a "stable" state. Growing evidences about the "microstructure" of NREM sleep present sleep as an ever-changing fluctuating state intermingled with microarousals, strongly contradicting to this simple model.

Keywords Hypothalamus • Sleep regulation • Sleep switch • Flip-flop model NREM sleep • Wake-sleep transition • Sleep microstructure

The first important step in understanding wake/sleep regulatory system in the brain was the discovery of the Viennese neurologist Baron Constantin von Economo in 1930. Based on clinicopathologic studies on victims of pandemic encephalitis lethargica, he observed that the prolonged sleepiness of one part of the victims is due to posterior hypothalamus and rostral midbrain injury. He has also observed that those affected people who suffered prolonged insomnia had lesions of the preoptic area and basal forebrain. Based on his observations, Economo assumed that the region of the hypothalamus near to the optic chiasm should contain sleep-promoting, while the posterior hypothalamus wake-promoting neurons (von Economo 1930). The key role of these two regions in sleep/wake regulation

P. Halász, R. Bódizs, *Dynamic Structure of NREM Sleep*,
DOI 10.1007/978-1-4471-4333-8_1,

has been confirmed by animal experiments (Nauta 1946; Swett and Hobson 1968; McGinty and Sterman 1968).

Almost 20 years later, Moruzzi and Magoun (1949) discovered the ascending reticular arousal system (RAS), originating in the upper brainstem, fuelled by the sensory brainstem input pathways. During the subsequent years, the different chemical transmission branches of RAS system were explored in detail (Jones 2003).

The question how this powerful system of wakefulness is switched of when we fall asleep remained enigmatic for a long time. At the turn of the twentieth/twenty-first century, Saper et al. (2001) and Gallopin et al. (2000) showed clearly that the ventrolateral preoptic area (VLPO and extended VPLO) sends GABA-ergic and galaninergic inhibitory impulses to all the brainstem nuclei harboring the ascending pathways of the arousal systems and keeps firing throughout the whole NREM sleep, providing the substrate of the "sleep system" having opposite function to the "wake system." Later, it turned out that the arousal systems also exert inhibitory effect on the "sleep-promoting" preoptic neurons (Saper et al. 2010).[1]

The discovery of the orexinergic system as lastly recognized special member of the arousal system provided further evidences for the mutually antagonistic system governing the actual balance between wakefulness and sleep (Nishino 2011). The area of basal forebrain has been thoroughly explored looking for sleep-promoting neuronal groups. It has been established by the c-Fos method that the ventrolateral preoptic nucleus (VLPO), a small circumscript area, exerts enduring neuronal activity during animal sleep. Further studies showed that in the preoptic region, a diffusely situated area (MnPN) has sleep-promoting activity too. Additional confirmation of the sleep-promoting role of this area came from neurotoxic destruction experiments of the VLPO and MnPN neurons (McGinty et al. 2004).

It has also been demonstrated that VLPO and the suprachiasmatic nucleus have synchronized activity, and both of them receive input from the retinal ganglionic cells. Therefore, circadian and visual information may modulate the VLPO activity that would be the substrate of circadian timing for the sleep process. Interesting tract-tracing and stimulation studies confirmed the presence of reciprocal inhibitory connection between the VLPO/MnPN neurons and the neurons of the ascending arousal system. These data serve as the substrate of the "sleep switch model" (Saper et al. 2001, 2010) based on the mutual inhibition between VLPO and the major arousal systems (Fig. 1.1).

> When VLPO neurons fire during sleep they would inhibit the arousal system cell groups thus disinhibiting and reinforcing their own firing. Similarly when arousal neurons fire at high rate during wakefulness, they would inhibit the VLPO, thereby disinhibiting their own firing. This reciprocal relationship is similar to a type of circuit that electrical engineers call 'flip-flop.' The two halves of a flip-flop circuit, by each strongly inhibit the other; create a feedback loop that is bistable, with two possible stable patterns of firing and a tendency to avoid intermediate states (Saper et al. 2001).

---

[1] Interestingly, certain neurons in the cholinergic nucleus basalis were found to be active during NREM sleep in association with cortical activation (Détári et al. 1984; Szymusiak and McGinty 1986; Détári and Vanderwolf 1987).

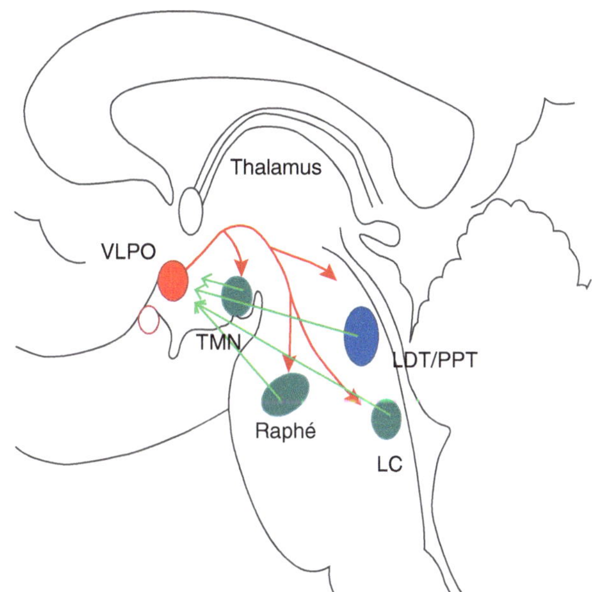

Fig. 1.1 A flip-flop switch model for wake-sleep regulation. Preoptic sleep-promoting neurons inhibit the arousal systems during sleep and are inhibited by the arousal systems during wake. As a result, the two mutually inhibitory systems form a flip-flop switch that is stable in either state but switches rapidly between them. *Red arrows* sleep promotion, *green arrows* wake promotion, *VLPO* ventrolateral preoptic area, *LDT/PPT* laterodorsal/ pedunculopontine tegmental cells, *LC* locus coeruleus, *TMN* tuberomammillary nucleus (Modified from Saper et al. 2001)

The experimental evidence for the sleep switch model of Saper is mainly based on the work of Takahashi and coworkers (2006, 2009, 2010). The Japanese group systematically detected the neuronal firing pattern of different sleep- and wake-promoting hypothalamic neuronal populations during falling asleep, sleeping, and awakening. Transitions from wake to sleep were characterized by the decrease of the firing rate of neurons of the wake-promoting and rising in the sleep-promoting neuronal network, and during awakening, an opposite dynamics of firing pattern was detectable (Fig. 1.2). These results confirm that these subcortical neuronal assemblies indeed have governing role in the EEG and behavioral changes during the sleep/wake cycle. However, the dynamics of falling asleep showed that the initiation of sleep is caused by the decreased activity of the wake-promoting neurons (disfacilitation) and not by the activity of sleep-promoting neurons (see also Steriade 2001). These results need confirmation and further research.

However, the course of sleep – depicted even roughly in the conventional hypnograms (Fig. 1.3) – shows us that sleep is not a big black hole which we fall in ("off switch") at evening and from which we suddenly get out ("on switch") at awakening in the morning. Sleep is organized in cycles where sleep goes deeper and deeper for a while, but after cc. 60 min, the direction of the process changes and stepwise became more and more superficial giving place to the first REM period. This sequence of event repeats in 4–6 times shaping the consecutive sleep cycles. In other words, within the NREM sleep period, robust dynamical changes take place, about which the flip-flop model has nothing to say about.

NREM sleep cannot be considered as a "stable" state, and as we will see in later parts of this book, indeed, growing evidences about the "microstructure" of NREM sleep (Terzano et al. 1985) present sleep as an ever-changing fluctuating

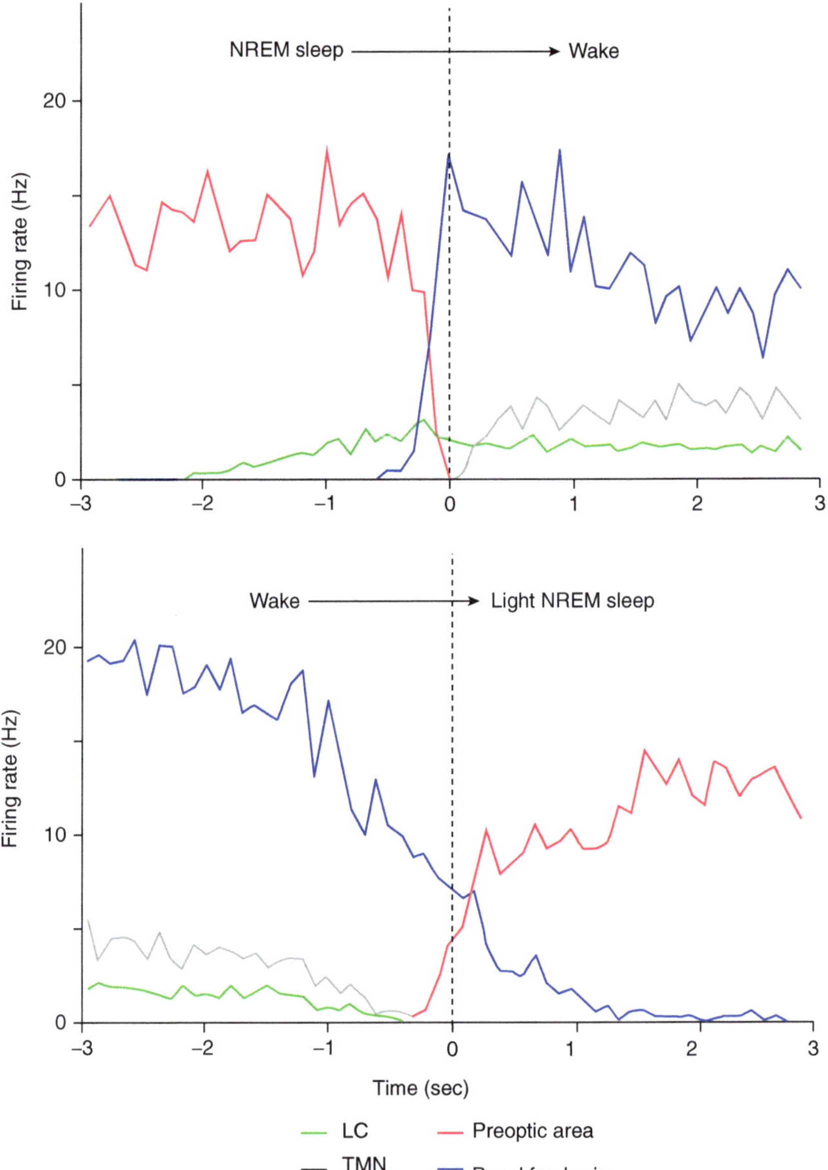

Fig. 1.2 Reciprocal firing patterns between sleep-promoting neurons in the preoptic area and wake-promoting neurons in the locus coeruleus (*LC*), tuberomammillary nucleus (*TMN*), and basal forebrain. The *top panel* shows changes in firing rate during the transition from NREM sleep to wake, and the *bottom panel* shows firing rates during the transition from wake into light NREM sleep. Note that the firing rates of some cell groups, such as the LC, begin to increase or decrease 1–2 s in advance of awakening or falling asleep, suggesting that they may help drive the transition. In contrast, neurons in the TMN begin to fire only after the transition to wake, suggesting these cells may play more of a role in maintenance of wakefulness. Recordings were made in unanesthetized, head-restrained mice (Adapted from Saper et al. (2010))

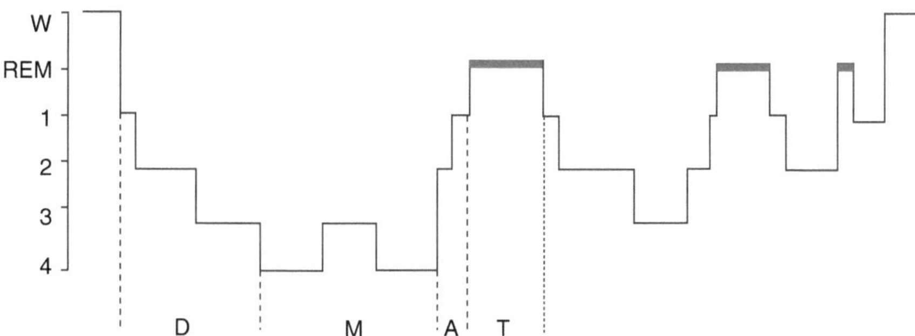

Fig. 1.3 The macrostructure of sleep as revealed by the conventional scoring system. Additional information on the descending (*D*), middle (*M*), ascending (*A*), and top (*T*) parts of a sleep cycle is given

state intermingled with microarousals, strongly contradicting this simple model (Halász et al. 2004).

Our guiding principle in this monography will be just the opposite view: sleep is essentially an unstable (multistable) state where the transitions from one state to another are not distinct, but gradual and the different levels of instability are maintained by continuous oscillation between sleep- and arousal-promoting influences. The macrotendencies prevail through oscillations: microstates build up macrostates. We will propose that input-dependent instability, expressed by phasic activation/deactivation, is an essential inbuilt feature of NREM sleep.

References

Détári L, Vanderwolf CH. Activity of identified cortically projecting and other basal forebrain neurones during large slow waves and cortical activation in anaesthetized rats. Brain Res. 1987; 437(1):1–8.

Détári L, Juhász G, Kukorelli T. Firing properties of cat basal forebrain neurones during sleep-wakefulness cycle. Electroencephalogr Clin Neurophysiol. 1984;58(4):362–8.

Gallopin T, Fort P, Eggermann E, Cauli B, Luppi PH, Rossier J, Audinat E, Mühlethaler M, Serafin M. Identification of sleep-promoting neurons in vitro. Nature. 2000;404(6781):992–5.

Halász P, Terzano M, Parrino L, Bódizs R. The nature of arousal in sleep. J Sleep Res. 2004;13(1): 1–23.

Jones BE. Arousal systems. Front Biosci. 2003;8:s438–51.

McGinty DJ, Sterman MB. Sleep suppression after basal forebrain lesions in the cat. Science. 1968,160(833):1253–5.

McGinty D, Gong H, Suntsova N, Alam MN, Methippara M, Guzman-Marin R, Szymusiak R. Sleep-promoting functions of the hypothalamic median preoptic nucleus: inhibition of arousal systems. Arch Ital Biol. 2004;142(4):501–9.

Moruzzi G, Magoun HW. Brain stem reticular formation and activation of the EEG. Electroencephalogr Clin Neurophysiol. 1949;1(4):455–73.

Nauta WJ. Hypothalamic regulation of sleep in rats; an experimental study. J Neurophysiol. 1946; 9:285–316.

Nishino S. Hypothalamus, hypocretins/orexin, and vigilance control. Handb Clin Neurol. 2011;99: 765–82.

Saper CB, Chou TC, Scammell TE. The sleep switch: hypothalamic control of sleep and wakefulness. Trends Neurosci. 2001;24(12):726–31.

Saper CB, Fuller PM, Pedersen NP, Lu J, Scammell TE. Sleep state switching. Neuron. 2010; 68(6):1023–42.

Steriade M. Active neocortical processes during quiescent sleep. Arch Ital Biol. 2001;139(1–2): 37–51.

Swett CP, Hobson JA. The effects of posterior hypothalamic lesions on behavioral and electrographic manifestations of sleep and waking in cats. Arch Ital Biol. 1968;106(3):283–93.

Szymusiak R, McGinty D. Sleep-related neuronal discharge in the basal forebrain of cats. Brain Res. 1986;370(1):82–92.

Takahashi K, Lin JS, Sakai K. Neuronal activity of histaminergic tuberomammillary neurons during wake-sleep states in the mouse. J Neurosci. 2006;26(40):10292–8.

Takahashi K, Lin JS, Sakai K. Characterization and mapping of sleep-waking specific neurons in the basal forebrain and preoptic hypothalamus in mice. Neuroscience. 2009;161(1):269–92.

Takahashi K, Kayama Y, Lin JS, Sakai K. Locus coeruleus neuronal activity during the sleep-waking cycle in mice. Neuroscience. 2010;169(3):1115–26.

Terzano MG, Mancia D, Salati MR, Costani G, Decembrino A, Parrino L. The cyclic alternating pattern as a physiologic component of normal NREM sleep. Sleep. 1985;8(2):137–45.

von Economo C. Sleep as a problem of localization. J Nerv Ment Dis. 1930;71(3):1–5.

Chapter 2
Dynamic NREM Sleep Regulation Models

Abstract In this chapter we provide an overview about those NREM sleep regulation models which take into consideration the effects of external and internal input factors apparently unrelated to the core sleep regulatory mechanisms but deeply influencing their dynamism. McCarley and Massaquoi (J Sleep Res 1(2):132–137, 1992) have begun to incorporate the influence of external noise according to the observations showing frequent brief nonbehavioral EEG and polygraphic "awakenings" in sleep. Lo et al. (Proc Natl Acad Sci USA 101(50): 17545–1758, 2004) studying brief sleep-wake transitions were able to show that these events can be commonly observed across different species with different sleep patterns. The universality of the distributions of short wake episodes strikingly contrasts the species-specific distributions of sleep bouts. Lo concludes that this relationship reveals a universal regulatory mechanism shaping the dynamism of sleep. Behn et al. (J Neurophysiol 97(6):3828–3840, 2007) created a model of sleep-wake network composed of coupled relaxation oscillation equations. This model could be considered as a crucial one in trying to explain the dual nature of sleep-regulation: gross sleep-wake regulatory mechanisms depending on the already described neural circuitry of the flip-flop switch and fine structure shaped by short bouts of wakefulness. We have hypothesized a parallel regulation of sleep in our model (Halász et al. J Sleep Res 13(1):1–23, 2004). Tonic processes were hypothesized to involve mainly intracerebral, slow, and chemical influences, while the phasic ones extracerebral, fast, and neuronal-synaptic ones tailoring the interaction of the reciprocal antagonistic influence between the sleep and arousal centers depicted in the flip-flop model of Saper et al. (Trends Neurosci. 24(12):726–731, 2001). The specificity of our model relied in the differential analysis of the descending and ascending slopes of the sleep cycles, which are usually undifferentiated in current models of sleep regulation.

Keywords Sleep regulation • Sleep-wake transition • Sleep micro-structure Power-law behavior • Coupled relaxation oscillators • Phasic regulation • Arousals in sleep Ascending slope • Descending slope

P. Halász, R. Bódizs, *Dynamic Structure of NREM Sleep*,
DOI 10.1007/978-1-4471-4333-8_2,

Previous models of sleep-wake behavior were based on network dynamics of neuronal systems assumed to having control over behavioral states (McCarley and Hobson 1975; Borbély 1982; Daan et al 1984). External input factors, according to the open system characteristics, were initially not taken into consideration. Later, Mc Carley has begun to incorporate the influence of external noise (McCarley and Massaquoi 1992) according to the observations showing frequent brief nonbehavioral EEG and polygraphic "awakenings" in sleep (Schieber et al. 1971; Halász 1982, 1993). These brief awakening-like phasic events were often discounted as "noise," and they have been only recently recognized as essential element of sleep architecture (Dijk and Kronauer 1999; Halász et al. 2004; Lo et al. 2004).

This combined approach: attributing significance not only to the network properties of wake- and sleep-promoting systems but also to external input aspects driving phasic events in the system has obtained strong support from studies using mathematical models for characterizing network dynamics behind sleep-wake behavior of experimental animals.

Lo et al. (2004) studying brief sleep-wake transitions were able to show that these events can be commonly observed across different species having different sleep patterns. They hypothesized that these brief awakenings from sleep may reflect aspects of endogenous sleep control mechanism. Analyzing sleep recordings from mice, rats, cats, and humans, they found that durations of the episodes during sleep exhibit a scale-free power-law behavior with an exponent ($\alpha = 2.2$) that remains the same for all investigated species. In contrast, sleep episode duration follows a species-specific exponential distribution, depending on body mass and metabolic rate. These findings indicate that brief awakenings from sleep are controlled by species-independent mechanisms in the sleep-wake neural networks. Lo et al. (2004) hypothesize that these could be determined by structural characteristics of the neural networks, the nature of the fluctuations around the sleep-wake transitional threshold, or other neurophysiologic features independent of species-specific body size. Furthermore, it is possible that internal and external inputs may excite wake-promoting neurons, leading to brief awakenings with power-law characteristics remaining the same across species. This dual, wake- and sleep-dependent regulation of the process resembles the dynamism known as self-organized criticality, determining recurring neural avalanches in the brain, emerging from quiet states. In coherence with this assumption, the size and duration of neural excitations in cortical networks were shown to follow power-law behavior, while quiet episodes follow a metabolic-rate-dependent timescale. In other words, the gross measures of sleep-regulation (sleep homeostasis and the ultradian cycles) follow the well-known rules of sleep regulation guided by the flip-flop circuitry, while on the finer timescale, this processes are shaped and scored by an alternative regulatory mechanism which is species independent finding its roots in some network properties, hitherto unrevealed by Lo et al. (2004).

Bohn et al. (2007) created a model of sleep wake network composed of coupled relaxation oscillation equations. This model could be considered as a crucial one in trying to explain the dual nature of sleep-regulation: gross sleep-wake regulatory mechanisms depending on the already described neural circuitry of the flip-flop switch and fine structure shaped by short bouts of wakefulness. The model is based on physiological data and can be considered as a further development of the concepts described by Lo et al. (2004). Mathematical analysis of the deterministic model provided insight into the dynamics underlying state transitions and predicted mechanisms for each transition time. With addition of noise, the stimulated sleep-wake behavior generated by the model reproduced many qualitative and quantitative features of mouse sleep-wake behavior. They wrote: "In contrast to the expected behavior of a pure flip-flop switch, this activity does not necessarily exhibit transition of the network from sleep to sustained wake: instead, switching between states depends on the strength of the impulse. If the pulse sufficiently depresses activity in sleep, the network will show transition from sleep to wake: otherwise the activity in wake self – terminates with falling phase of the oscillation and sleep bout continues." The strength of the pulse inhibition from wake to sleep during a brief awakening is terminated by a variable describing homeostatic NREM sleep drive. In general, depending on the strength of inhibition from sleep to wake or the reverse, the effect of stimulation could be influenced by the state dominance of sleep- or wake-promoting forces on the descending or ascending slopes of the cycles, as we describe later.

The Behn paper emphasizes that their model predicts conditions determining concurrent activity in wake and sleep-active neuronal populations. This is specifically the case when awakening impulses arrive in a state of high homeostatic drive, and no sustained transition to wake state occurs. Transcending the all-or-nothing type approach of the original flip-flop model, without denying its basic tenets and strengths, is the major advantage of this model. However, neither the Lo et al. (2004) nor the Behn et al. (2007) paper considers the difference between the descending and ascending slopes of the sleep cycles which seems to be a crucial one in sleep regulation.

We have hypothesized a parallel regulation of sleep in our model (Halász et al. 2004). Tonic processes were hypothesized to involve mainly intracerebral, slow, and chemical influences, while the phasic ones extracerebral, fast, and neuronal-synaptic ones tailoring the interaction of the reciprocal antagonistic influence between the sleep and arousal centers depicted in the flip-flop model (Saper et al. 2001, 2010). This parallel working mode of the tonic and phasic processes provides the flexibility of sleep maintaining at the same time the core features of endogenously determined, species-specific sleep structure. The specificity of our model relied in the differential analysis of the descending and ascending slopes of the sleep cycles, which are usually undifferentiated in current models of sleep regulation (Fig. 2.1). In our opinion, this is a particularly important issue in the dynamics of ultradian sleep cycles. Examples are the differential distribution of the microstructural elements, as well as the differences in cardiocerebral coupling along the descending and ascending slopes. Taking into consideration the striking sleep-microstructural differences between these slopes, we suggested that the descending slope is characterized by an increasing influence of the sleep system exerting inhibition over the wake system, while the opposite trends were attributed to the ascending slopes. The dynamism of these influences determine the sensorial inputs from the

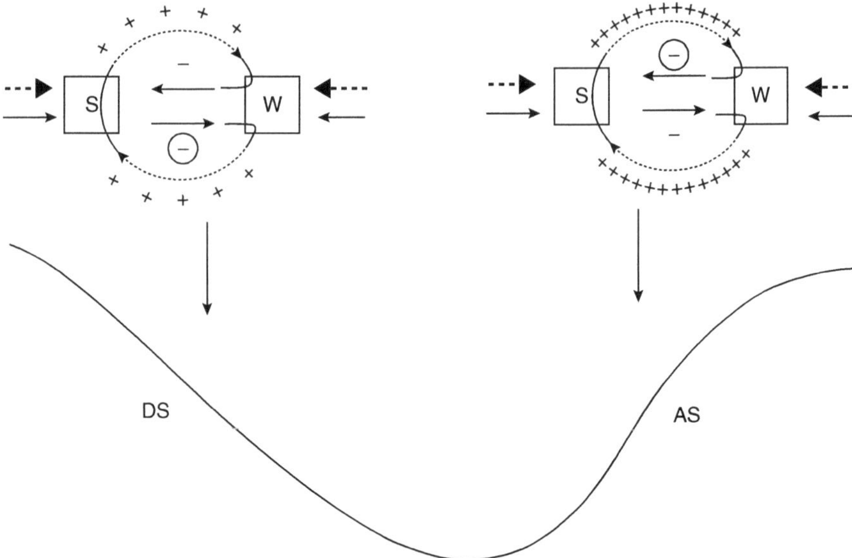

Fig 2.1 Assumed tonic and phasic interaction between sleep (*S*)- and wake (*W*)-promoting neuronal assemblies during the descending (*DS*) and ascending (*AS*) slopes of sleep cycles. *Arrows* (with positive and negative signs) between the S and W boxes indicate reciprocal antagonistic relationship between sleep- and wake-promoting neuronal groups. Descending slope is characterized by dominance of sleep promotion, while ascending slope by wake promotion. When sleep promoting forces are overwhelming (on the descending slope), the external and internal sensorial input (*thick interrupted arrow*) is rare, and the elicited responses appear in sleep like (K-complexes, slow waves) form. Contrasting on the ascending slope, the inhibition of the wake-promotion neurons decrease, and the sensorial input is able to elicit more frequent arousal like responses. *Continuous arrows* indicate the endogenous chemical influences, *intererupted arrows* the external sensorial influences.

internal and external environment to evoke mild and infrequent arousals in the descending slopes of the sleep cycles, while during the ascending slopes, a higher frequency and intensity of arousals is facilitated by the increasing influence of the wake system. The arousals contribute to the sleep regulatory process by fuelling the dynamics of the sleep cycles potentially resulting in sleep facilitation during the descending slope and awakening in the ascending one (Halász et al. 2004). The sleep-promoting effects of sensory stimulation (Bohlin 1971; Oswald 1960; Webb and Agnew 1979) as well as the deepening of sleep during the descending slopes of its cycles in presence of sensory stimulation (Hirshkowitz 2002) were reported but remained largely unexplained and unmodeled in the literature. Although innovative and integrative, our arousal model did not directly address the issue of sleep homeostasis, which is crucial in understanding the nature of sleep. Some sorts of microarousals or reactive sleep EEG elements are composed by the slow EEG waves known to be the physiological basis of sleep homeostasis. Thus, the question arises whether these reactive slow elements contribute to the process of sleep homeostasis or are some different processes resembling but not contributing to it? In other words, what is the relationship between sleep homeostasis and the microstructure of sleep?

In the following chapters of this book, we will base our reasoning on observations about the microstructural dynamics of NREM sleep and experimental data on the functional characteristics of wake- and sleep-promoting subcortical networks.

References

Behn CG, Brown EN, Scammell TE, Kopell NJ. Mathematical model of network dynamics governing mouse sleep-wake behavior. J Neurophysiol. 2007;97(6):3828–40.

Bohlin G. Monotonous stimulation, sleep onset and habituation of the orienting reaction. Electroencephalogr Clin Neurophysiol. 1971;31(6):593–601.

Borbély AA. A two process model of sleep regulation. Hum Neurobiol. 1982;1(3):195–204.

Daan S, Beersma DG. Borbély AATiming of human sleep: recovery process gated by a circadian pacemaker Am J Physiol. 1984;246(2 Pt 2):R161–83.

Dijk DJ, Kronauer RE. Commentary: models of sleep regulation: successes and continuing challenges. J Biol Rhythms. 1999;14(6):569–73.

Halász P. Arousals without awakening – dynamic aspect of sleep. Physiol Behav. 1993;54(4): 795–802.

Halász P. The role of the non-specific sensory activation in sleep regulation and in the pathomechanism of generalized epilepsy with generalized spike-wave discharge. Doctoral thesis, the Hungarian Academy of Sciences, Budapest; 1982.

Halász P, Terzano M, Parrino L, Bódizs R. The nature of arousal in sleep. J Sleep Res. 2004; 13(1):1–23.

Hirshkowitz M. Arousals and anti-arousals. Sleep Med. 2002;3(3):203–4.

Lo CC, Chou T, Penzel T, Scammell TE, Strecker RE, Stanley HE, et al. Common scale-invariant patterns of sleep-wake transitions across mammalian species. Proc Natl Acad Sci USA. 2004; 101(50):17545–8.

McCarley RW, Hobson JA. Neuronal excitability modulation over the sleep cycle: a structural and mathematical model. Science. 1975;189(4196):58–60.

McCarley RW, Massaquoi SG. Neurobiological structure of the revised limit cycle reciprocal interaction model of REM cycle control. J Sleep Res. 1992;1(2):132–7.

Oswald I. Falling asleep open-eyed during intense rhythmic stimulation. Br Med J. 1960; 1(5184):1450–5.

Saper CB, Chou TC, Scammell TE. The sleep switch: hypothalamic control of sleep and wakefulness. Trends Neurosci. 2001;24(12):726–31.

Saper CB, Fuller PM, Pedersen NP, Lu J, Scammell TE. Sleep state switching. Neuron. 2010; 68(6):1023–42.

Schieber JP, Muzet A, Ferriere PJ. Phases of spontaneous transitory activation during normal sleep in humans. Arch Sci Physiol (Paris). 1971;25(4):443–65.

Webb WB, Agnew Jr HW. Sleep onset facilitation by tones. Sleep. 1979;1(3):281–6.

Chapter 3
Recognition of Spontaneous and Evoked Arousal- and Sleep-Like (Antiarousal) Phasic Events

Abstract In this chapter, we will deal with data accumulated about the arousal-driven spontaneous and evoked phasic events. We are looking for characteristics of spontaneous and elicited phasic changes, regularities in their occurrence, and the impact of their existence on the course of sleep. Summarizing the results of earlier works about PAT, K-complexes, and CAP phenomena, a new approach has been used in their systematization. First, we have differentiated arousal-like and sleep-like spontaneous and elicited responses and investigated the relation of these two types with dynamic changes in the course of night sleep. We differentiated these phasic changes from tonic states of sleep and analyzed the interrelationships of the phasic and tonic changes.

Keywords Phases d'activation transitoire • K-complexes • Micro-arousals • Reactive delta waves • Cyclic alternating pattern

Dijk and Kronauer wrote in a comment on "Models of sleep regulation: successes and continuing challenges" in 1999 that "although sleep models may describe adequately global sleep patterns and their circadian modulation, detailed modelling of the frequent short awakenings from, and the subsequent transitions back to sleep, as well as the variation of the propensity to awaken across the ultradian non-REM-REM cycle, is not addressed. Incorporation of these aspects of sleep in mathematical models of sleep regulation may improve further our understanding of key aspect of sleep regulation, that is, its timing" (Dijk and Kronauer 1999).

DOI 10.1007/978-1-4471-4333-8_3,

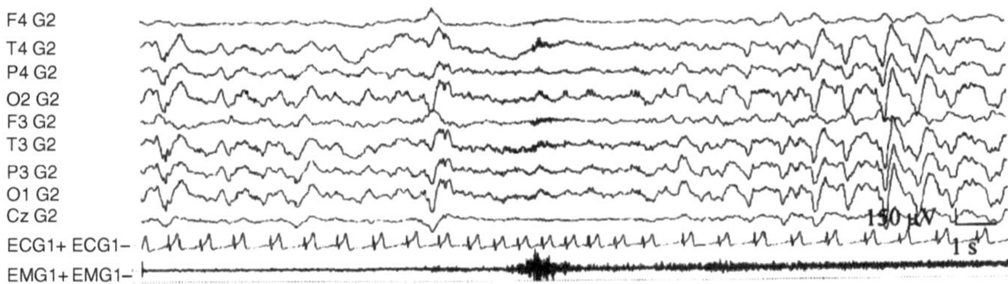

Fig. 3.1 Example of a spontaneous PAT phenomenon during NREM sleep. Sleep EEG (stage 3) became suddenly desynchronized with appearance of faster rhythms concomitant with transitory increase in muscle activity and heart rate. After the phasic event, a rebound-like increase in EEG slow wave activity can be seen parallel with transitory heart rate decrease

3.1 "Phases d'Activation Transitoire" (PAT) (First Recognition that Arousal-Like Events Are Standard Constituents of NREM Sleep)

The phenomenon of spontaneous recurrent arousals with characteristic EEG changes and polygraphic signs, without behavioral awakening was first described Schieber et al. (1971) named at that time as "phases d'activation transitoire" (PAT). The criteria for PAT in NREM sleep given by Schieber et al. (1971) were the following: "Increase in EEG frequencies in conjunction with decrease of amplitudes, disappearance of delta waves and spindles, transitory enhancement of muscle tone or phasic appearance of groups of muscle potentials, movements of the limbs or changes in body posture, transitory rise in heart rate."

The duration of these changes varied from some seconds to more than 10 s. Temporary activation is followed by deactivation leading to a biphasic character of the phenomenon (Fig. 3.1). The term was modified by several workers in the last years and used in the context of physiologic and pathologic studies, with more or less the same meaning and criteria (Quattrochi et al. 2000; Sforza et al. 2002).

The occurrence of PAT negatively correlates with the depth of sleep, occurring more frequently in superficial than in deep sleep with the highest incidence during REM sleep and stage 1 and appearing less frequently during stage 3 and 4 (Fig. 3.2). Erhart and Muzet (1974) in the same research group have studied the elicitability of PAT by acoustic and thermal stimuli. They had the fundamental finding on the existence of the same arousal pattern (evoked PATs) in response to artificial stimuli as they appear spontaneously. They have registered beside the evoked PATs spontaneous PATs as well but in diminished number compared to the nights without stimulation. Therefore, they assumed a control mechanism regulating the number of arousals during the night sleep.

PAT is the first systematically studied phasic event identified with arousal/activation recurring during sleep. It has been suggested by the Brussels scholars, that spontaneous activation periods are related to endogenous impulses of the reticular activating system.

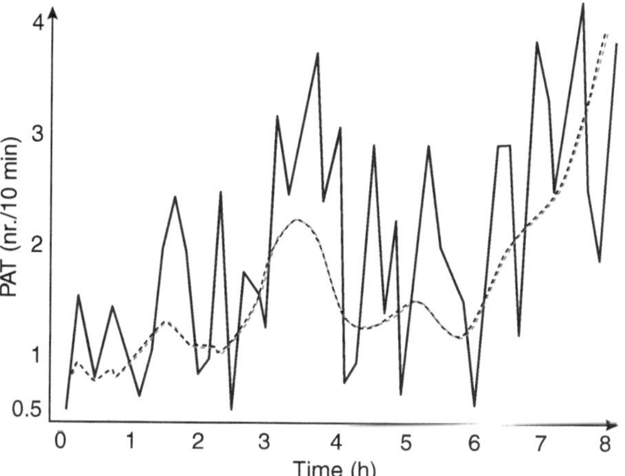

Fig. 3.2 The evolution of PAT (phases d'activation transitoire) events during a night sleep in ten young healthy volunteers (After Schieber et al. 1971)

3.2 K-Complexes: The First Hit in Recognition of Reactive Deltas in NREM Sleep

The first observation proving the elicitability of delta waves was the discovery of K-complexes (Blake and Gerard 1937; Loomis et al. 1939; Davis et al. 1939). Researchers were already fascinated by the resemblance of the spontaneously occurring isolated slow waves to those ones they were able to induce by external stimuli delivered during sleep. Two recent reviews are available to have a broader overview beyond the scope of this book (Colrain 2005; Halász 2005).

K-complexes during sleep appear at 5 months of age. K-complexes are best distinguishable in stage 2 sleep, as a spontaneously occurring phasic events, but it is possible to elicit them by sensory stimuli as well. In deeper stages, high-voltage slow waves practically absorb K-complexes but with averaging, the same kinds of K-complex-like waves have been elicited in stage 3–4 sleep (Ujszászi and Halász 1988; Bastien and Campbell 1992; Niiyama et al. 1995). The frequency of spontaneous K-complexes is between 1 and 2/min in most studies (Halász et al. 1985).

The components of the K-complex are in close relationship to or can be identified with late components of sensory evoked potentials in sleep, as shown in studies on K-complexes evoked by acoustic stimuli (Ujszászi and Halász 1988; Bastien and Campbell 1992; Niiyama et al. 1995; Riedner et al. 2011). Components of the potential complex that occur in response to short tone pips in NREM sleep stages 2–3 are generally accepted as follows: N100, P200 (or 250), N300 (or 350), P400, N550, P900, N1500, and P1900 (Fig. 3.3). It was proved that N350 could be present even when N550 and P900 do not follow or are absent (Bastien and Campbell 1992) and held to be identical with the "vertex sharp wave."

The first studies of K-complexes showed that they are elicitable by all modalities of sensory stimuli (Bastien and Campbell 1992; Niiyama et al. 1995; Roth et al. 1956; Sallinen et al. 1994), but most easily by acoustic stimuli and are

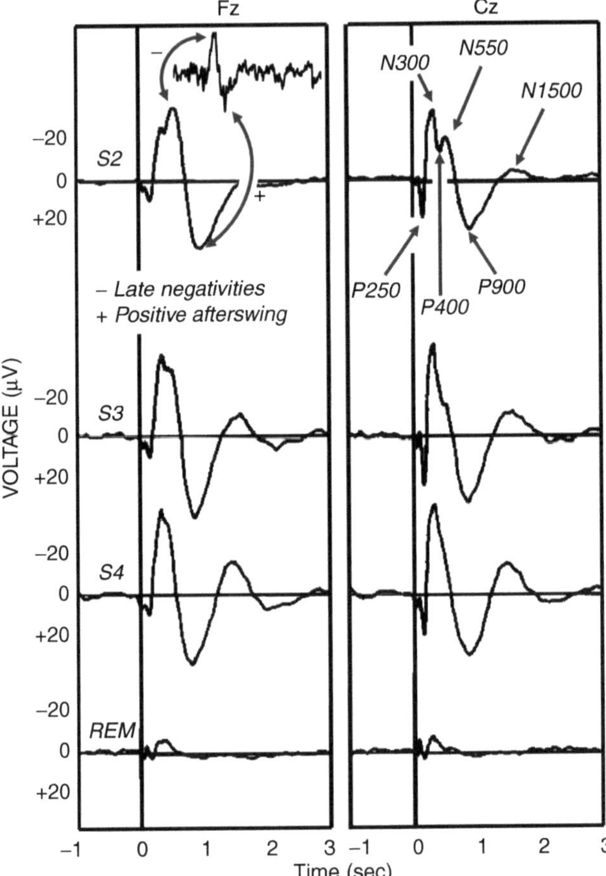

Fig. 3.3 K-complexes as auditory evoked potentials in different stages of human sleep reflected in Fz and Cz electrodes. A spontaneous singular event sharing the main features of the evoked and averaged events (late negative components and positive afterswing) is shown in the *top left* of the figure. Components of the auditory evoked potentials are defined on the *top right*. Note the persistence of the main features and the similarity between spontaneous and evoked K-complexes. *S2* stage 2, *S3* stage 3, *S4* stage 4, *REM* rapid eye movement sleep

accompanied by autonomic discharges identical to those seen for arousals (Ackner and Pampiglione 1957; Fruhstorfer 1971, 1995; Hornyak et al. 1991; Johnson and Karpan 1968; Sassin and Johnson 1968; Sforza et al. 2000, 2002; Takigawa et al. 1980).

The distribution and amplitudes of the late components of the K-complex proper (N350-N550-P900) vary according to three factors: topography, level of sleep, and information content (novelty and meaning) of eliciting stimuli (Ujszászi and Halász 1988). The whole complex is not homogeneous but probably represents a cascade of events originating from sources of different topography that are mobilized by different processes. They are activated in a certain order provoked by the nature and context of the eliciting stimuli (Halász 1993).

Continuously or periodically applied random acoustic stimulation increased the number of K-complexes and simultaneously decreased the number of the spontaneous K-complexes (Halász 1982; Halász et al. 1985). (This issue will be elaborated in Chap. 4.)

While several data clearly show that K-complexes have arousal response properties, other studies indicate that K-complexes are "fore runners" of NREM delta activity, their amplitude increase after sleep deprivation, similarly to deltas (De Gennaro et al. 2000; Nicholas et al. 2002; Peszka and Harsh 2002), and they are held to be sleep-promoting or preserving agents (Wauquier et al. 1995). The distribution of K-complexes from cycle to cycle across the night proved to show parallel course with homeostatic decay of slow waves (Halász et al. 1985; Rajna et al. 1983).

We have emphasized repeatedly that K-complexes have "Janus-faced" features based on their arousal dynamic contrasted with their sleep-maintaining, anti-arousal properties (Halász et al. 1985; Halász 2005).

Bastuji et al. (1995) developed the forced awakening method. In this paradigm, subjects were questioned about quantitative and qualitative aspects of stimulus recall evoked by oddball-type stimuli in parallel with recording of the evoked cortical responses, after being aroused by the stimuli from naps. In subjects whose quality of recall was excellent, P300 waves were indistinguishable from those obtained before sleep. When P300 was found attenuated, delayed, and desynchronized, recall was quantitatively degraded and P300 was concomitant to or replaced by sleep negativities (varieties of late negative components being part of the K-complex) in subjects in whom stimulus recall was severely degraded or absent. They concluded that K-complex-analogue sleep negativities have two aspects being, on the one hand, arousal driven and, on the other, erasers preventing accurate memory encoding and retrieval of the stimulus, consequently promoting sleep. Thus, the same Janus-faced nature of K-complexes was stressed by them similarly to our previous work.

Earlier studies showed already that K-complexes and delta waves have prefrontal-frontal localization emanating from a wide bilateral field (Ujszászi and Halász 1988; Bastuji et al. 1995; Colrain et al. 2000; Bastien et al. 2002). The scalp topography of K-complexes evoked by acoustic and respiratory stimuli was similar (Gora et al. 2001). Also, the large positive afterswing like P900 component reached an amplitude-maximum over the fronto-central areas (Ujszászi and Halász 1988; Bastien and Campbell 1994; Cote et al. 1999) and the frontal dominance became more prominent with deepening of sleep to stages 3–4 NREM. The frontal maximum of delta waves during deep sleep (stages 3–4) was evident long ago and is still supported by several power spectra studies of NREM sleep (Horne 1993; Cajochen et al. 1999; Finelli et al. 2001; Marzano et al. 2010).

Czisch and coworkers (2009) using EEG-fMRI have found maximal activation during sound tone evoked K-complexes in the middle frontal gyri and cingulate areas. In another recent study, Reidner et al. (2011) applied a more sophisticated EEG source analysis with sLORETA method and confirmed that the bilateral anterior cingulate, middle frontal, inferior frontal orbital, and rectal gyri were the most pronounced areas in terms of relative current during K-complexes evoked by different (auditory, somatosensory, and visual) stimulation. Furthermore, they were able to detect a modality-specific activation (by relatively increased current) in specific cortical areas in accordance with the modality of the applied type of stimulation at the 550 ms slow negative peak of the evoked K-complex.

These results provide a direct link between modality-specific sensory information processing and diffuse nonspecific K-complex-like slow wave responses during NREM sleep.

Recently, Jahnke et al. (2012) found in another fMRI analytic study found a wide spontaneous K-complex-associated network reflected by BOLD positive signal changes in subcortical (brain stem, thalamus), cerebellar, sensory (auditory and visual), motor midline (anterior and midcingulate gyrus, precuneus), and other regions which form part of the default mode network. They emphasized that the primary auditory cortex was the first cortical region to be influenced during the K-complex. Their interpretation is very much the same as in our earlier works:

K-complex embodies an arousal with subsequent sleep guarding counteraction that might on one hand serve monitoring of the environment with basic information processing and on the other hand protect continuity of sleep and thus its restoring effect.

In another recent paper, Koshaka et al. (2012) reported on transient activation of the ventral brain stem preceding the K-complex by detection of auditory brain stem-evoked responses and a sustained activation of the dorsal brain stem outlasting the K-complex. Thus, it is suggested that K-complexes are triggered by the activation of the brain stem.

The question how primary sensory pathways are able to activate in NREM sleep the nonspecific slow wave response is presently unknown. One possibility would be the parallel activation of the thalamocortical system producing slow waves when arousal system is shut down. Another assumption can be the secondary involvement of larger cortical fields via cortico-cortical connection after the initial cortical modality-specific activation. A further approach is outlined in Sect. 6.2.

Intracranial distribution of K-complexes obtained by stereo-electroencephalographic recordings has been detected recently. Congruently with the scalp topography and source analytic studies the intracranial distribution of K-complexes has been found to be maximal over the anterior medial and superior frontal cortices (Wennberg 2010).

A very important new line in the K-complex research has been started by the report of Amzica and Steriade (1997) showing the relation of K-complexes and slow oscillation of NREM sleep below 1 Hz and later by Cash et al. (2009) pointing out a particular relationship with the large negative main component of K-complex and the down state of the slow oscillation (Fig 3.4).

In stage 2 sleep, 18–20 % of K-complexes are followed by long lasting changes in the ongoing EEG. They are accompanied by other rhythms such as K-delta, K-alpha, and K-spindle according to the nature of the associated rhythm (Raynal et al. 1974; Halász and Ujszászi 1991; MacFarlane et al. 1996).

Although isolated K-complexes are present in stage 2 mainly at the more superficial level, the majority of them are part of a longer phasic event called by us as "micro-arousals." The term micro-arousal for these changes was used to designate those phasic EEG events which were not associated with awakenings

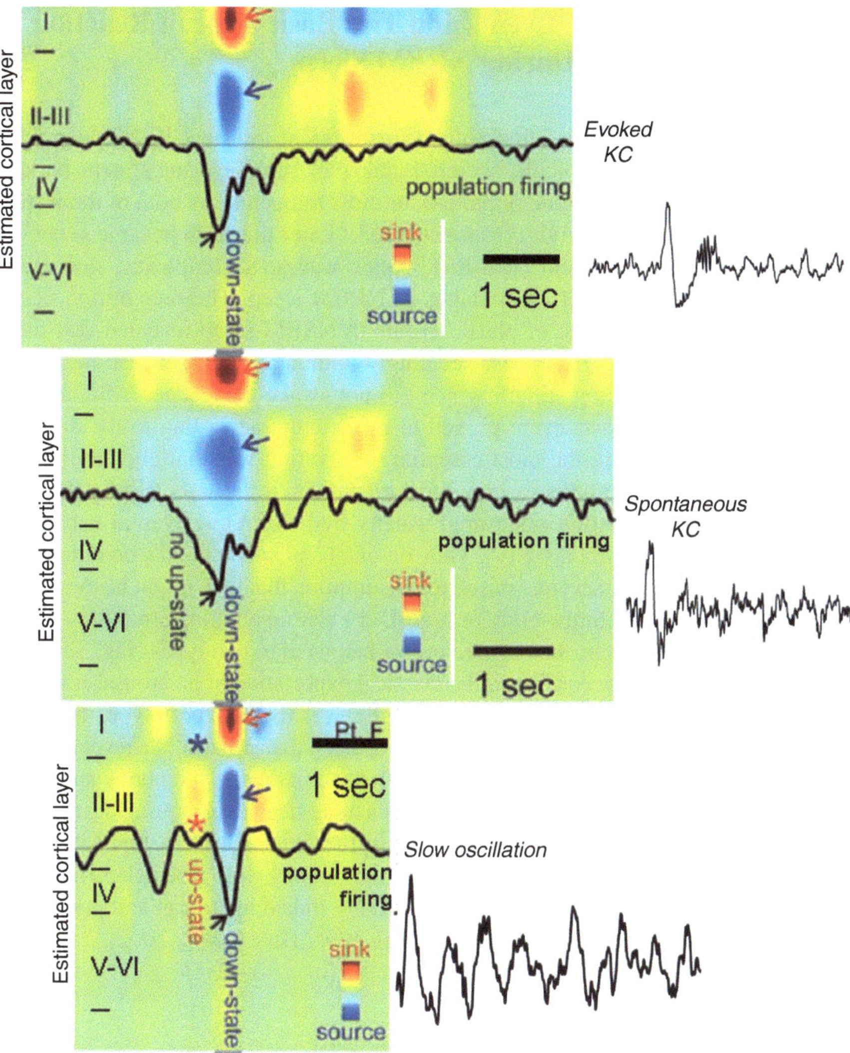

Fig. 3.4 The cortical down state as a common denominator of K-complexes and slow waves. Evoked and spontaneous K-complexes of stage 2 sleep are common in being based on singular down states, while the down state of the slow oscillation of stages 3–4 is embedded in alternating up-down state sequences. The figure is the result of coregistered multiunit activity and current source density with an intracortical multimicroelectrode in human subjects. *Red arrow* indicates inward currents (sink) in superficial cortical layer and *blue arrows* show outward currents (source) in layers II–III. *Black arrow* indicates decreased neuronal firing (Modified from Cash et al. 2009)

regardless of their desynchronizational (fast, low amplitude) or synchronizational (sleep response-like) morphology and regardless of their connection with autonomic phenomena or some sort of behavioral arousal (Halász et al. 1979).

3.3 Nature of Phasic Events and Two Basic Type of Reaction to Phasic Input During NREM Sleep

The term "phasic event" was first coined for different transient REM sleep EEG phenomena (like ponto-geniculo-occipital spikes or rapid eye movement bursts), which differ either in amplitude or pattern, or most frequently by both of them from the ongoing background activity (Moruzzi 1963). Later this term became generally used for other transient event (like PAT, vertex waves, K-complexes, slow wave groups) observed during different stages of NREM sleep. The recognition of the universal recurring presence of phasic events in NREM sleep revealed that these events, although they might appear seemingly without any eliciting stimulus, are elicitable by mild arousing stimuli which did not awake the sleeper. Although the potential to evoke a phasic event proved to be different according to the different modalities of the stimuli, the most essential observation was that elicitability was not modality specific. Another essential feature of them is the association with more or less autonomic activation (measured usually by heart rate elevation and motor response/increase of EMG activity). This, on one hand spontaneous, on other hand reactive feature of phasic events, raised the assumption that they might be evoked by unnoticed internal (within the body or brain), or external eliciting stimuli.

Whatever the input was, the reactive events belonged to two types (Fig. 3.5). One group of them behaved according to the classical expectation of an arousal response: increase of frequency and decrease of amplitude. It is the PAT prototype. The other group of phasic events was characterized by mor sleep than arousal like response with K-complexes, and slow groups, without or with lower intensity autonomic and motor response, the EEG answer was characterized by a more sleep than arousal like response with K-complexes, slow wave groups. Therefore, the "arousal-like" nature of this type was from the beginning questioned raising debates among sleep researchers about the nature of this Janus-faced reactivity being partially linked to arousal and partially to sleep induction or maintenance (Wauquier et al. 1995; Hirshkowitz 2002).

3.4 Consequences of the Recognition Phasic Events with Contrasting Features During NREM Sleep

During the 1970s and 1980s of the last century, it became clear that level of sleep continuously oscillates in every stage of NREM sleep, and these oscillations are related to phasic events. First, the activation phases (PAT) and K-complexes were recognized as singular events, later it became evident that they are associated with more complex and longer changes. We started to call them as micro-arousals (Halász et al. 1979). Therefore, underlying the more or less stereotype sleep process which runs its course every night, there is another layer constituted by phasic element, representing different forms of micro-arousals, being less stereotype and more shaped by the sensorial input, influencing the individual and actual course of sleep.

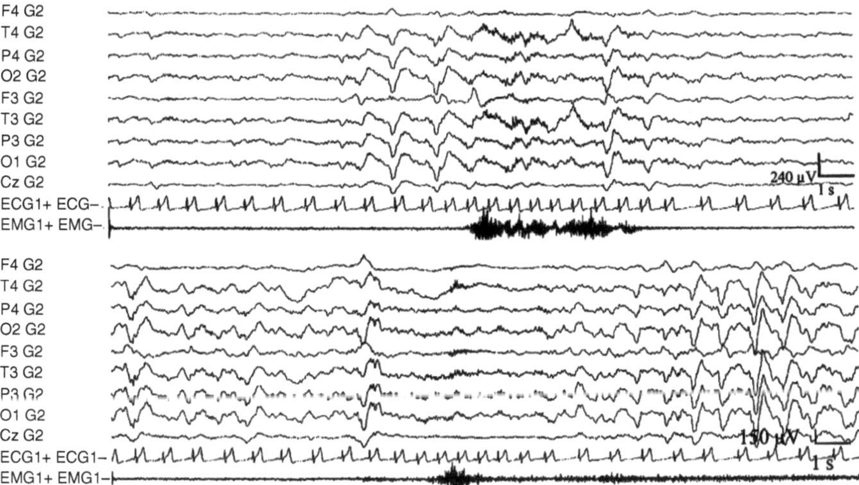

Fig. 3.5 Characteristic examples of synchronizational (**a**) and desynchronizational (**b**) phasic changes during NREM sleep. The essential contrasting features are seen in the EEG parts: in (**a**) a sudden transitory increase of slow wave activity, while in (**b**) the transitory appearance of fast rhythms suppressing the ongoing slow wave activity can be observed. Simultaneous autonomic and motor activation are almost the same during both responses

From this time, a new dynamic view emerged, and the thinking about the nature of arousals has changed: they were no more held as simple random disturbance signals being harmful for sleep but as something which is more organically woven into the texture of sleep, participating in the regulation of sleep. Researchers who became interested in the exploration of the dynamic microstructure of sleep bellow the sleep stages of the Rechtschaffen-Kales staging system started to organize workshops and the material of the *Second International Workshop on Phasic Events and Dynamic Organization of Sleep* was published in 1991 (Terzano et al. 1991).

Among the seemingly activation/arousal-driven micro-events, an interesting paradoxical group has emerged which was associated with more sleep than arousal-like appearance (called later as antiarousal). It gained more and more importance and lastly led to the recognition of the cyclic alternating pattern (CAP).

3.5 Recognition of Cyclic Alternating Pattern (CAP): A System Reflecting the Degree of NREM Sleep Instability

This line of research related to phasic events comes from the studies of the Parma group who discovered the CAP phenomenon in the mid-1980s (Terzano et al. 1985). This turned out to be a general framework for all the dynamics of phasic events and micro-arousals described in previous studies.

The CAP is an EEG activity of NREM sleep characterized by sequences of transient electrocortical events that are distinct from background EEG activity and recur at up to 1-min interval (Terzano et al. 1985). The CAP cycle consists of two phases: a phase A ("activation") and a phase B ("background") (Fig. 3.6). Sensory stimuli applied during phase B are able to elicit the phase A pattern. In CAP, the input-dependent phasic events are arranged in complex pseudo-periodic assemblies. The mean period time between two A phases is 20–40 s. The average phase A duration is 10–12 s, while it is 20–30 s for phase B.

In a detailed analysis of CAP sequences and non-CAP periods (Smerieri et al. 2007) the mean duration of CAP sequences was 2 min and 33 s, and their number varied between 28 and 54 with an average of approximately 44/night. Each CAP sequence was composed of an average of 5.6 CAP cycles. The length of CAP sequences proved to be determined by the number (and not by duration) of CAP cycles. They defined CAP cycles as "strings of time constant modules involved in tailoring sleep structure."

According to the type of activation reached by phase A, three categories (Fig. 3.7) have been differentiated (Terzano and Parrino 2000).

Phase A_1 type comprises exclusively synchronization patterns (alpha in stage 1, sequential K-complexes in stage 2 and superficial stage 3, and reactive slow wave sequences in stages 3 and 4). It is identical with the aforementioned synchronization (or sleep like)-type micro-arousals or explicitly antiarousals.

Phase A_2 type is composed of micro-arousals preceded by synchronization composed of K-complexes followed by other components (sigma, alpha, and delta stretches).

On the highest level of arousal, the phase A_3 type will be a micro-arousal without synchronization. This is identical with the PAT pattern of the Strasbourg group or the arousal pattern recognized by ASDA (1992).

The percentage of CAP time in NREM sleep (CAP rate) is age related (Parrino et al. 2012), increasing from preschool age (25.9 %) to peripubertal age (62.1 %) then slightly decreasing in the middle age (37.5 %) and again increasing importantly in senium (55.3 %). If we analyze the CAP rate according to the three subtypes, the peribupertal peak can be attributed to the rate of A_1 type, while the peak in senescence comes from an increase of A_2–A_3 rate. Estimating CAP rate in 1–4 months' age, the big majority of A_1 types (85.2 %) is conspicuous. In summary, more than 80 % rate of the A_1 type was seen in infancy, school, and peripubertal ages, which are milestones of mental development. The course along lifetime shows opposite trends for A_1 versus A_2–A_3 type A phases: while A_1 rate decreases A_2–A_3 rate increases. The opposite trends converge around the age of 60 (Fig. 3.8). The duration of CAP cycles (length from A to the next A phase) seems to be rather stable.

The CAP rate correlates negatively with the subjective evaluation of the quality of sleep: the higher the CAP rate, the poorer the quality of sleep (Terzano and Parrino 2000). The CAP rate is also increased by external noise and lowered by prolonged sleep deprivation. Stimulating and arousing drugs increase, and hypnotic/sedative drugs decrease the CAP rate.

Topographic mapping studies of the CAP revealed that spectral components with anterior frontal prevalence was found in A_1 type (0.25–2.5 Hz) events, while the A_3

type (7–12 Hz) events have different prevalence over the parietooccipital areas (Fig. 3.9). The low-frequency components (A_1) were expressed more close to the midline, and the higher frequency bands (A_3) were more distributed over the parietooccipital convexity (Ferri et al. 2005a). Also, the highest synchronization of the two hemispheres emerged from the electrode pair F3–F4 (Ferri et al. 2005b, 2006a).

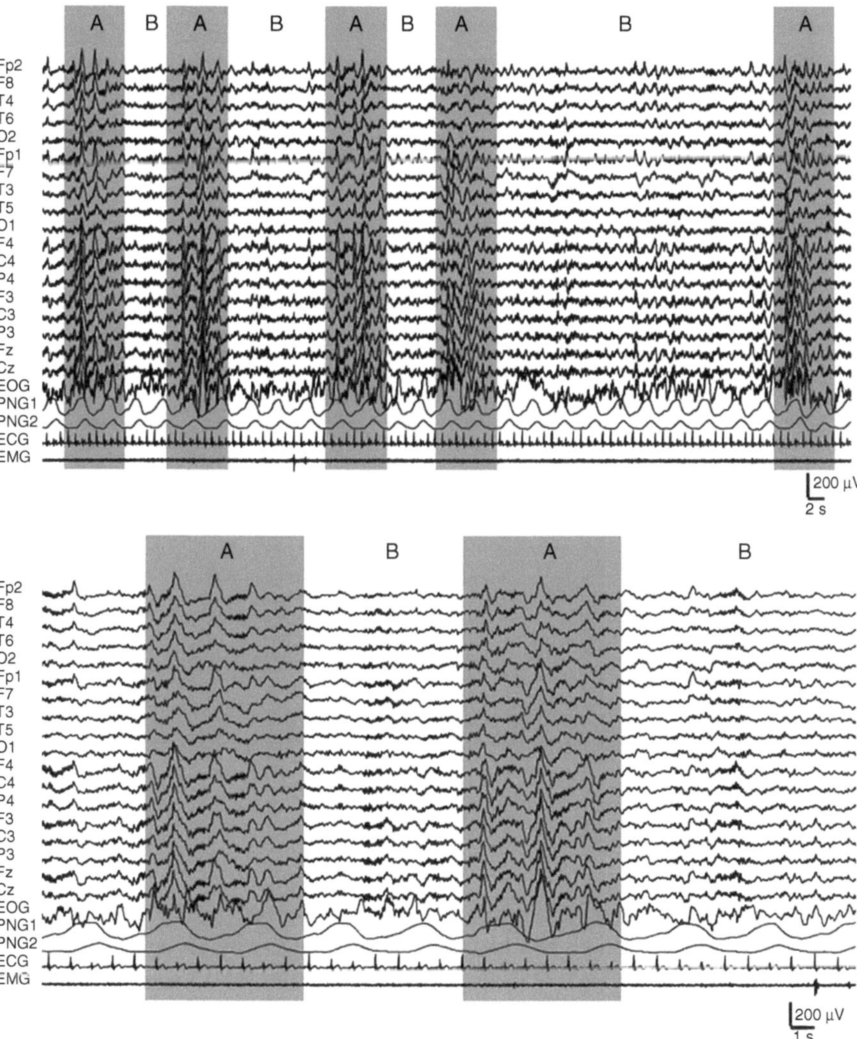

Fig. 3.6 Example for the phenomenon of cyclic alternating pattern (*CAP*) during sleep. The first plate is a longer period highlighting the amplitude variations of NREM sleep EEG slow wave activity. The following plates are successive zooms of the first one, showing the alternation of *A* (activity) and *B* (background) phases during the CAP sequence. Details of the record are 26-year-old healthy women and contralateral mastoid reference

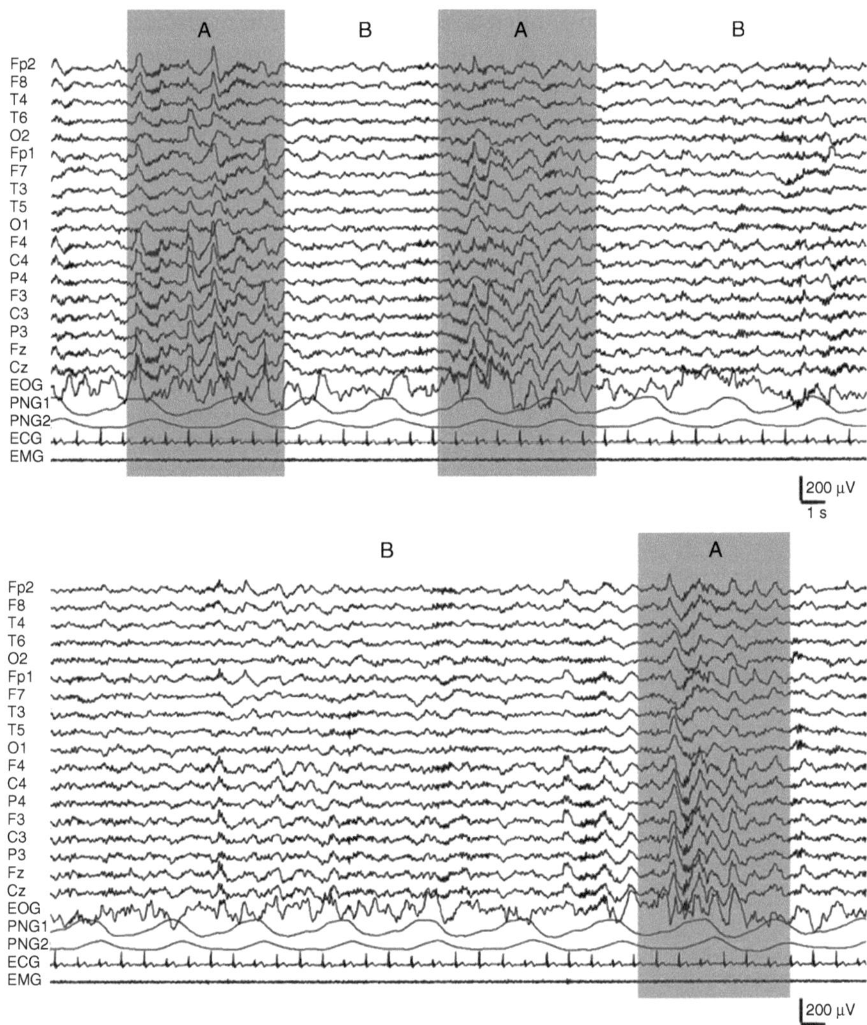

Fig 3.6 (continued)

The power spectral analysis of CAP phenomena showed (Ferri et al. 2005a) that CAP corresponds to periods in which very slow delta activity groups together with a range of different EEG activities with higher (alpha and beta) frequencies. CAP B phase proved to contain less sigma activity compared to NCAP periods.

The concept of CAP offers a global framework for measuring sleep instability (Terzano and Parrino 2000; Parrino et al. 2012). The appearance of CAP sequences represents an arousal control mechanism. They reflect to all of the arousing influences and set into motion an oscillatory level-setting system around the referential state. They provide a flexible adaptation for the system to defend against perturbations.

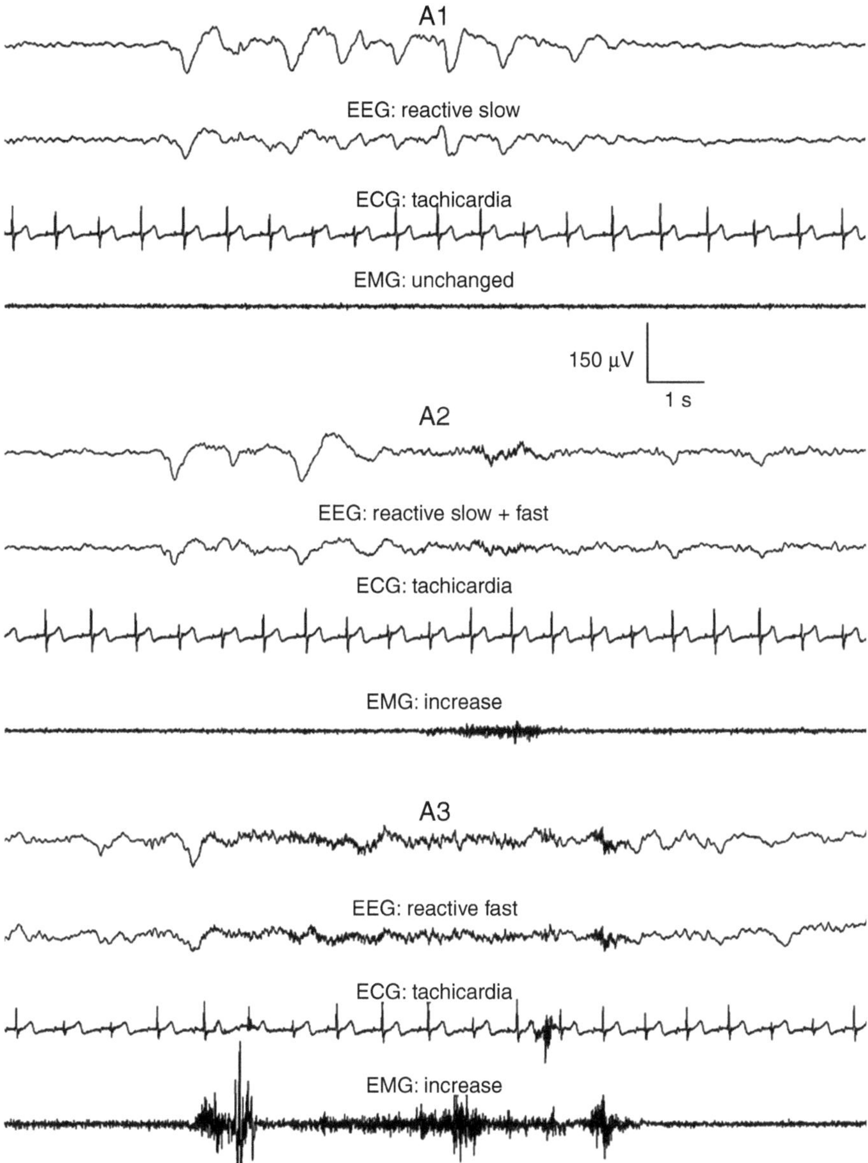

Fig. 3.7 Representative examples of the three types of CAP-A-phases. Details in the text. Calibration refers to the EEG traces

The time structure of CAP sequences was recently analyzed in more details by Ferri and coworkers (Ferri et al. 2006b). Using the Markovian analysis as approach, shorter than 60 s (peak between 20 and 40 s) interval was characteristic during slow wave sleep of adults between the A phases, while the intervals

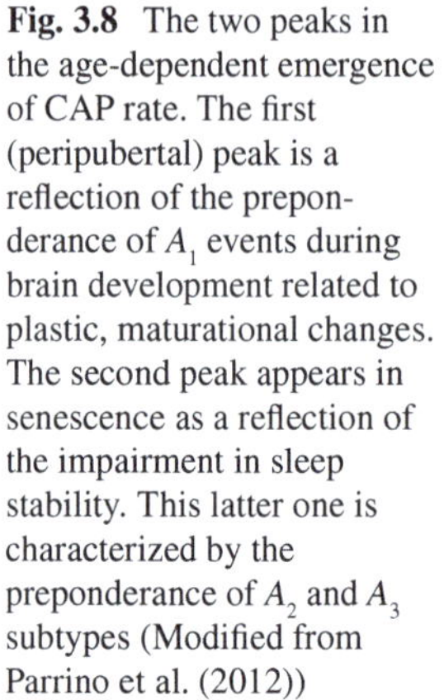

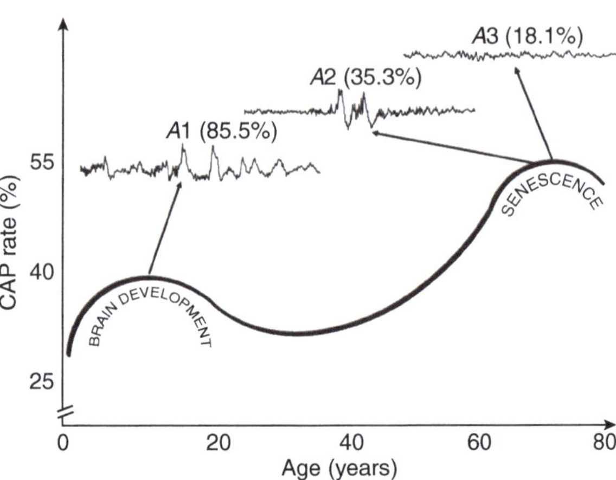

Fig. 3.8 The two peaks in the age-dependent emergence of CAP rate. The first (peripubertal) peak is a reflection of the preponderance of A_1 events during brain development related to plastic, maturational changes. The second peak appears in senescence as a reflection of the impairment in sleep stability. This latter one is characterized by the preponderance of A_2 and A_3 subtypes (Modified from Parrino et al. (2012))

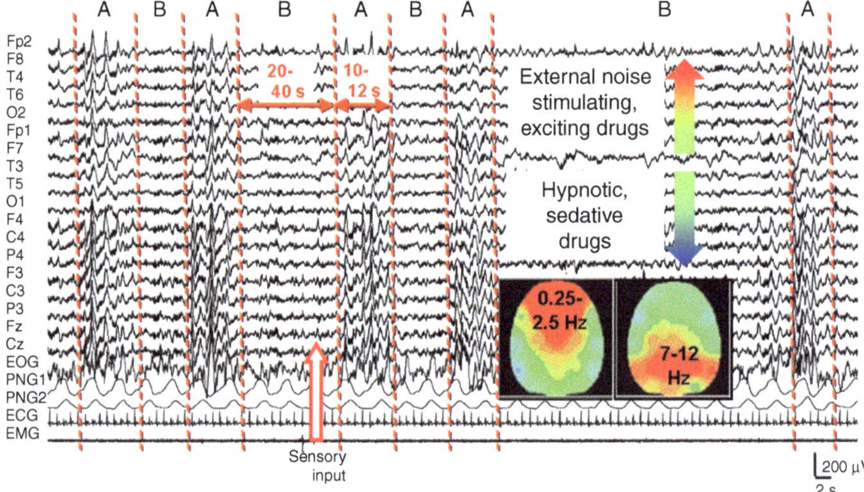

Fig. 3.9 Basic parameters and modulating factors of CAP. Average interval between two A phases and the mean duration of the A phases are shown. The topography and power spectra of A_1 and $A_{2/3}$ types is different: A_1 is dominated by frontal midline 0.25–2.5 Hz activity, while $A_{2/3}$ by 7–12 Hz posterior lateral activity

were longer during more superficial sleep. The serial repetition of CAP phases, or in other words the transitional probability of CAP sequences, proved to be not random but deterministic with lower values of entropy and higher values of time dependency in slow wave sleep. Presumably the increased length of CAP sequences plays a buffer effect against perturbing factors, which induces increase of instability in sleep but reduces sleep fragmentation at the same time (Parrino et al. 2012).

3.6 Recognition of NREM Sleep Microstructure Conceived in CAP: Theoretical and Practical Values

Studies on the CAP system – conducted in overwhelming majority by the Parma group and collaborators – have utmost importance in understanding how sleep is adapted by this level setting, self-regulatory device to incoming impulses, using up these impulses to maintain or even build up sleep. The differences among the three groups of CAP A phases (A_1–A_2–A_3) indicate the actual dynamic balance between the oscillations of sleep- and wake-promoting forces, probably originating in the reciprocal antagonistic brain stem networks determining cortical responsivity. A_3 type micro-arousal means an actual wake promoting preponderance, A_1 indicates preponderance of sleep promoting forces, whereas A_2 reflects an in-between state when sleep promoting preponderance is weak and a breakthrough of wake promotion happens.

Parallel with their physiological significance, the measurable parameters (e.g., CAP rate) of the system may show us the degree of sleep instability and therefore became ideal markers of different sleep pathologies (Parrino et al. 2012).

References

Ackner B, Pampiglione G. Some relationships between peripheral vasomotor and E.E.G. changes. J Neurol Neurosurg Psychiatry. 1957;20(1):58–64.

Amzica F, Steriade M. The K-complex: its slow (<1-Hz) rhythmicity and relation to delta waves. Neurology. 1997;49(4):952–9.

ASDA (American Sleep Disorders Association). EEG arousals: scoring rules and examples. Sleep. 1992;15:173–84.

Bastien C, Campbell K. The evoked K-complex: all-or-none phenomenon? Sleep. 1992;15(3): 236–45.

Bastien C, Campbell K. Effects of rate of tone-pip stimulation on the evoked K-complex. J Sleep Res. 1994;3(2):65–72.

Bastien CH, Crowley KE, Colrain IM. Evoked potential components unique to non-REM sleep: relationship to evoked K-complexes and vertex sharp waves. Int J Psychophysiol. 2002;46(3): 257–74.

Bastuji H, García-Larrea L, Franc C, Mauguière F. Brain processing of stimulus deviance during slow-wave and paradoxical sleep: a study of human auditory evoked responses using the oddball paradigm. J Clin Neurophysiol. 1995;12(2):155–67.

Blake H, Gerard RW. Brain potentials during sleep. Am J Physiol. 1937;119:692–703.

Cajochen C, Foy R, Dijk DJ. Frontal predominance of a relative increase in sleep delta and theta EEG activity after sleep loss in humans. Sleep Res Online. 1999;2(3):65–9.

Cash SS, Halgren E, Dehghani N, Rossetti AO, Thesen T, Wang C, et al. The human K-complex represents an isolated cortical down-state. Science. 2009;324(5930):1084–7.

Colrain IM. The K-complex: a 7-decade history. Sleep. 2005;28(2):255–73.

Colrain IM, Webster KE, Hirst G, Campbell KB. The roles of vertex sharp waves and K-complexes in the generation of N300 in auditory and respiratory-related evoked potentials during early stage 2 NREM sleep. Sleep. 2000;23(1):97–106.

Cote KA, de Lugt DR, Langley SD, Campbell KB. Scalp topography of the auditory evoked K-complex in stage 2 and slow wave sleep. J Sleep Res. 1999;8(4):263–72.

Czisch M, Wehrle R, Stiegler A, Peters H, Andrade K, Holsboer F, Sämann PG. Acoustic oddball during NREM sleep: a combined EEG/fMRI study. PLoS One. 2009;4(8):e6749.

Davis H, Davis PA, Loomis AL, Harvey EN, Hobart G. Electrical reactions of the human brain to auditory stimulation during sleep. J Neurophysiol. 1939;2:500–14.

De Gennaro L, Ferrara M, Bertini M. The spontaneous K-complex during stage 2 sleep: is it the 'forerunner' of delta waves? Neurosci Lett. 2000;291(1):41–3.

Dijk DJ, Kronauer RE. Commentary: models of sleep regulation: successes and continuing challenges. J Biol Rhythms. 1999;14(6):569–73.

Ehrhart J, Muzet A. Frequency and duration of transitory activation phases during normal or disturbed sleep in man. Arch Sci Physiol (Paris). 1974;28(3):213–60.

Ferri R, Bruni O, Miano S, Terzano MG. Topographic mapping of the spectral components of the cyclic alternating pattern (CAP). Sleep Med. 2005a;6(1):29–36.

Ferri R, Rundo F, Bruni O, Terzano MG, Stam CJ. Dynamics of the EEG slow-wave synchronization during sleep. Clin Neurophysiol. 2005b;116(12):2783–95.

Ferri R, Rundo F, Bruni O, Terzano MG, Stam CJ. Regional scalp EEG slow-wave synchronization during sleep cyclic alternating pattern A1 subtypes. Neurosci Lett. 2006a;404(3):352–7.

Ferri R, Bruni O, Miano S, Plazzi G, Spruyt K, Gozal D, et al. The time structure of the cyclic alternating pattern during sleep. Sleep. 2006b;29(5):693–9.

Finelli LA, Borbély AA, Achermann P. Functional topography of the human nonREM sleep electroencephalogram. Eur J Neurosci. 2001;13(12):2282–90.

Fruhstorfer H. Habituation and dishabituation of the human vertex response. Electroencephalogr Clin Neurophysiol. 1971;30(4):306–12.

Gora J, Colrain IM, Trinder J. The investigation of K-complex and vertex sharp wave activity in response to mid-inspiratory occlusions and complete obstructions to breathing during NREM sleep. Sleep. 2001;24(1):81–9.

Halász P. Arousals without awakening – dynamic aspect of sleep. Physiol Behav. 1993;54(4): 795–802.

Halász P. K-complex, a reactive EEG graphoelement of NREM sleep: an old chap in a new garment. Sleep Med Rev. 2005;9(5):391–412.

Halász P, Ujszászi J. Spectral features of evoked microarousals. In: Terzano MG, Halász P, Decklerck AC, editors. Phasic events and dynamic organization of sleep. New York: Raven Press; 1991. p. 85–100.

Halász P, Kundra O, Rajna P, Pál I, Vargha M. Micro-arousals during nocturnal sleep. Acta Physiol Acad Sci Hung. 1979;54(1):1–12.

Halász P. The role of the non-specific sensory activation in sleep regulation and in the pathomecha nism of generalized epilepsy with generalized spike-wave discharge. Doctoral thesis, Hungarian Academy of Sciences, Budapest; 1982.

Halász P, Pál I, Rajna P. K-complex formation of the EEG in sleep. A survey and new examinations. Acta Physiol Hung. 1985;65(1):3–35.

Hirshkowitz M. Arousals and anti-arousals. Sleep Med. 2002;3(3):203–4.

Horne JA. Human sleep, sleep loss and behaviour: implications for the prefrontal cortex and psychiatric disorder. Br J Psychiatry. 1993;162:413–9.

Hornyak M, Cejnar M, Elam M, Matousek M, Wallin BG. Sympathetic muscle nerve activity during sleep in man. Brain. 1991;114(Pt 3):1281–95.

Jahnke K, von Wegner F, Morzelewski A, Borisov S, Maischein M, Steinmetz H, et al. To wake or not to wake? The two-sided nature of the human K-complex. Neuroimage. 2012;59(2):1631–8.

Johnson LC, Karpan WE. Autonomic correlates of the spontaneous K-complex. Psychophysiology. 1968;4(4):444–52.

Kohsaka S, Sakai T, Kohsaka M, Fukudac N, Ariga A. Activation of the brainstem precedes and outlasts the K-complex in humans. Neuroscience. 2012;202:243–51.

Loomis AL, Harvey EN, Hobart GA. Distribution of disturbance-patterns in the human electroencephalogram, with special reference to sleep. J Neurophysiol. 1939;2:413–30.

MacFarlane JG, Shahal B, Mously C, Moldofsky H. Periodic K-alpha sleep EEG activity and periodic limb movements during sleep: comparisons of clinical features and sleep parameters. Sleep. 1996;19(3):200–4.

Marzano C, Ferrara M, Curcio G, De Gennaro L. The effects of sleep deprivation in humans: topographical electroencephalogram changes in non-rapid eye movement (NREM) sleep versus REM sleep. J Sleep Res. 2010;19(2):260–8.

Moruzzi G. Active processes in the brain stem during sleep. Harvey Lect. 1963;58:233–97.

Nicholas CL, Trinder J, Colrain IM. Increased production of evoked and spontaneous K-complexes following a night of fragmented sleep. Sleep. 2002;25:882–7.

Niiyama Y, Fushimi M, Sekine A, Hishikawa Y. K-complex evoked in NREM sleep is accompanied by a slow negative potential related to cognitive process. Electroencephalogr Clin Neurophysiol. 1995;95(1):27–33.

Parrino L, Ferri R, Bruni O, Terzano MG. Cyclic alternating pattern (CAP): the marker of sleep instability. Sleep Med Rev. 2012;16:27–45.

Peszka J, Harsh J. Effect of sleep deprivation on NREM sleep ERPs and related activity at sleep onset. Int J Psychophysiol. 2002;46(3):275–86.

Quattrochi JJ, Shapiro J, Verrier RL, Hobson JA. Transient cardiorespiratory events during NREM sleep: a feline model for human microarousals. J Sleep Res. 2000;9:185–91.

Rajna P, Halász P, Kundra O, Pál I. Event-related non-specific responses (K-complexes) during sleep. Acta Med Hung. 1983;40(1):33–40.

Raynal D, et al. K-alpha events in hypersomniacs and normals. Sleep Res. 1974;3:144.

Riedner BA, Hulse BK, Murphy MJ, Ferrarelli F, Tononi G. Temporal dynamics of cortical sources underlying spontaneous and peripherally evoked slow waves. Prog Brain Res. 2011;193:201–18.

Roth M, Shaw J, Green J. The form voltage distribution and physiological significance of the K-complex. Electroencephalogr Clin Neurophysiol. 1956;8(3):385–402.

Sallinen M, Kaartinen J, Lyytinen H. Is the appearance of mismatch negativity during stage 2 sleep related to the elicitation of K-complex? Electroencephalogr Clin Neurophysiol. 1994;91(2):140–8.

Sassin JF, Johnson LC. Body motility during sleep and its relation to the K-complex. Exp Neurol. 1968;22(1):133–44.

Schieber JP, Muzet A, Ferriere PJ. Phases of spontaneous transitory activation during normal sleep in humans. Arch Sci Physiol (Paris). 1971;25(4):443–65.

Sforza E, Jouny C, Ibanez V. Cardiac activation during arousal in humans: further evidence for hierarchy in the arousal response. Clin Neurophysiol. 2000;111(9):1611–9.

Sforza E, Juony C, Ibanez V. Time-dependent variation in cerebral and autonomic activity during periodic leg movements in sleep: implications for arousal mechanisms. Clin Neurophysiol. 2002;113(6):883–91.

Smerieri A, Parrino L, Agosti M, Ferri R, Terzano MG. Cyclic alternating pattern sequences and non-cyclic alternating pattern periods in human sleep. Clin Neurophysiol. 2007;118(10): 2305–13.

Takigawa M, Uchida T, Matsumoto K. Correlation between occurrences of spontaneous K-complex and the two physiological rhythms of cardiac and respiratory cycles (author's transl). No To Shinkei. 1980;32(2):127–33.

Terzano MG, Parrino L. Origin and significance of the cyclic alternating pattern (CAP). Sleep Med Rev. 2000;4(1):101–23.

Terzano MG, Mancia D, Salati MR, Costani G, Decembrino A, Parrino L. The cyclic alternating pattern as a physiologic component of normal NREM sleep. Sleep. 1985;8(2):137–45.

Terzano MG, Halász P, Declerck AC, editors. Phasic events and dynamic organization of sleep. New York: Raven Press; 1991.

Ujszászi J, Halász P. Long latency evoked potential components in human slow wave sleep. Electroencephalogr Clin Neurophysiol. 1988;69(6):516–22.

Wauquier A, Aloe L, Declerck A. K-complexes: are they signs of arousal or sleep protective? J Sleep Res. 1995;4(3):138–43.

Wennberg R. Intracranial cortical localization of the human K-complex. Clin Neurophysiol. 2010; 121(8):1176–86.

Chapter 4
The Cyclic Structure of Sleep: Relationship Between the Macrostructural Slopes of Cycles and Microstructural Dynamics

Abstract We are dealing with the relationship between macrostructural (slope of the sleep cycles) and the microstructural dynamics. Descending ("D") and ascending ("A") slopes of cycles show opposite trends: "D" slope shows sleep promoting, while "A" slope shows wake/REM-promoting tendencies. Arousal-like phasic changes (PAT and CAP A_{2-3} phases) turned out to be overwhelming during the "A" slopes of sleep cycles, and their amount is increased from evening to morning. Within the "A" slopes their frequency increases before REM sleep periods. Sleep-like phasic changes have a different course. They are prevalent during the "D" slopes of the first sleep cycles and follow the homeostatic decay of slow wave oscillation. Accordingly, the elicitability of the two types of phasic events has opposite tendencies during night sleep from evening to morning: Arousal-like responses are prevalent during the "A" slopes and their responsivity increases, while sleep-like responses are prevalent during the "D" slopes and their elicitability decreases. Lastly, we point out the double nature of the forces behind tonic and phasic regulation: There are alternating chemical influences behind the tonic one and faster reciprocal sleep- and wake-promoting inputs behind the phasic regulation. Due to the intermingled phasic and tonic regulation, the sleeper is connected with the surrounding world and is able to execute two seemingly contradictory tasks: (1) separation for trophotropic (use dependent homeostatic) functions and (2) reactivity in form of state-dependent responses.

Keywords Sleep cycles • Descending slopes • Ascending slopes • Microstructural dynamics • State-dependent reactivity • Tonic- and phasic-regulation of sleep

P. Halász, R. Bódizs, *Dynamic Structure of NREM Sleep*,
DOI 10.1007/978-1-4471-4333-8_4,

4.1 Opposite Trends Across Descending and Ascending Slopes of Sleep Cycles

Since the beginning of the contemporary sleep studies, the stereotype sequences of sleep stages during sleep cycles have become well-known. The majority of studies have engaged in the characteristics of NREM-REM alternations, and the rules governing the stage sequences within the NREM cycle have received less attention. In the 1970s, researchers started to study how the sequence of NREM sleep stages influences the NREM-REM cyclicity (Brezinová 1974). One of the most important trends was to analyze the process-like behavior of the cyclic events (Caille and Bassano 1976), for example, how the preceding events or sequences determine the consecutive stages. These studies have found evidence supporting the "semi-Markov" nature of sleep stage sequences where the appearance of a certain stage depends partially on the preceding stages and partially on timing within the sleep process. Along these studies, it has become obvious that the sleep process is not only a continuous repetition of certain sequence of stages, but other long term processes from evening to morning also take place: The deepness of the cycles (the amount of slow wave sleep) decreases, while the length and density of REM periods grow from evening to morning.

However, very few studies have been devoted to look for the mechanism behind the increasing-decreasing course of sleep depth during a sleep cycle. Blake and Gerard (1937) were the first who mentioned the increasing and decreasing states and EEG differences between them. They have emphasized that decreasing and increasing states are not like mirror images. Williams et al. (1964) analyzing stage sequences in sleep of young adult persons showed that on the deepening part of sleep cycles, the stages follow an order according to their depth, while on the A part of the cycles, some stages are skipped, and this part is usually shorter compared to the D one. Sinha et al. (1972) used the asymmetry (differences in steepness) of cycle slopes to measure the progress in the sleep process and forecast the time of awakening. In an early work (Halász 1982), we studied the order of sleep stages in two human sleep experiments according to their distribution along the "D" or "A" part of the cycle. The first two cycles of six whole night sleep records of one young healthy volunteer and in a second experiment whole night sleep records of eight other volunteers were studied. Part "D" was defined as a period of sleep between the waking state/REM sleep level and the first deepest phase of the given cycle, while part "A" as those periods of sleep between the last deepest phase end and the next highest level stage in the cycle. Our results are summarized in Fig. 4.1. The sequence of stages clearly showed the expected downward and upward order of stages along the "D" and "A" part of cycles, respectively. This was supported in 94.3 and 98.5 % of stage transitions in the 6 sleep records of the same subject and in the group of eight subjects, respectively. As a next step then, we measured the duration of the "D" and "A" slope of cycles in the same sleeps. A significant difference between the duration of "D" and A slopes has been found in the six sleeps of our young volunteers. The average duration of slope "D" was 37.7 min, while the duration of the slope

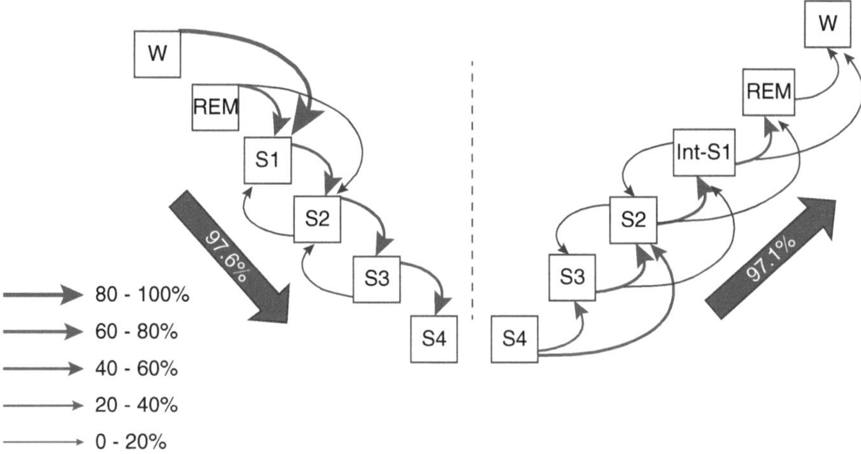

Fig. 4.1 Distribution of stage transitions during the descending and ascending slopes of sleep cycle in sleeps of eight healthy young volunteers. The *thickness of the arrows* indicates the occurrence (in percentage) of the indicated kind of transition. The *large arrow* shows that both on descending and ascending slopes of the cycle the overwhelming majority of stage transitions were going during the descending slope downward while during the ascending slope upward (Modified from Halász (1982))

A was only 21.9 min ($p<0.01$). The same kind of asymmetry has been found in the other eight volunteers. The number of stage shifts was also measured in the two groups. It proved to be significantly higher on the "D" slope, compared with the "A" slope. Looking for the stage jumps, when the transition skipped a stage in the depth-order of stages, the "A" slope contained more of them, and the A slopes were found to be more steep just because of the higher number of these "jumps."

Therefore, we can say that the two slopes of a certain sleep cycle reflect two different directions (downward vs. upward) of the sleep process. It seems probable that while the working mode of the "D" slope is "sleep promotion," it is "wake/REM promotion" for the "A" slope. From a dynamic point of view, they deeply differ from each other. The consequence of this consideration means that the seemingly homogenous stage levels along different slopes of a cycle are not homogenous, for example, descending stage 2 is probably different from ascending stage 2 in several functional aspects reflecting the actual balance between sleep- and wake-promoting forces.

Another approach was taken by Merica and Fortune (1997) following the dynamic changes within a NREM cycle by power spectrum analysis. Spectral analysis provided a quantitative measure and showed how much the different oscillations (slow waves, spindles, fast frequency oscillations, etc.) coexist at all times in sleep. Their ratio defines stage assignment within the cycles. There are two important aspects in this work. One is the clear demonstration that transition from waking to sleep and further stage transitions within NREM sleep, NREM-REM transitions as well as the awakening process are all not sudden events but step by step progressive procedures, each involving transitional periods. Importantly, they also tried to trace back the EEG changes to the reciprocal inhibitory dynamics on the cellular activity level of the sleep- and wake-promoting structures. This analysis clearly demonstrates that sleep is never a static state during none of the stages, either on the "D" or the "A" slopes of the cycles. They measured the simultaneous power-time courses during the cycles and demonstrated the characteristic composition pattern of delta waves, sigma spindles, and beta activity shaping the less steep "D" and steeper "A" slopes of cycles. This time course of the different frequencies may be the result of the progressive decrease in the firing rate of wake-promoting neurons. It results in diminished excitatory input to the thalamus and a consequent, gradual hyperpolarization of the thalamic relay cells, building up the burst-firing pattern, leading first to spindling and later to develop delta activity. Interestingly, during the "D" slope of cycles they have found multiple delta peaks, instead of one monotonic rise of deltas, and they assumed a correlation of this phenomenon with CAP events, otherwise not taken into consideration in their work.

The structure of sleep cycles, at least that of the first three ones is obviously shaped by an asymmetric (longer) "D" and (shorter) "A" slope. The "turning point" from deep sleep during the first cycles is very characteristic and frequently marked by a sudden movement. (This period is "abused" by pathological arousal events in children with pavor nocturnus and adults with night terror). The dynamics of this period and subsequent uprising course of sleep leading to REM sleep need further explanation.

4.2 Distribution of Phasic Events and Micro-arousals Across the Sleep Process

4.2.1 The Distribution of PAT-Like Micro-arousals Across the Sleep Process

The distribution of PAT is not homogeneous across the sleep cycles: They are more frequent during the A slopes of the cycles compared with the D slopes, and their frequency increases from evening to morning (Schieber et al. 1971). These early findings were confirmed by Halász (1982) and also later after coining the American Sleep Disorders Association (ASDA) criteria of arousal (ASDA 1992; Terzano et al. 2002). The distribution of arousals consistent with ASDA criteria has followed similar distribution during night sleep: increasing from evening to morning (Terzano et al. 2002).

4.2.2 The Distribution and Dynamics of K-Complexes Across the Sleep Process

Regardless of the level of sleep or sleep stage, the number of K-complexes seems to correlate with the actual direction of sleep process. In our earlier studies (Halász 1982), the number of K-complexes during the stage 2 sleep periods was rather variable. When the "D" types (preceded by a more superficial sleep stage and followed by a

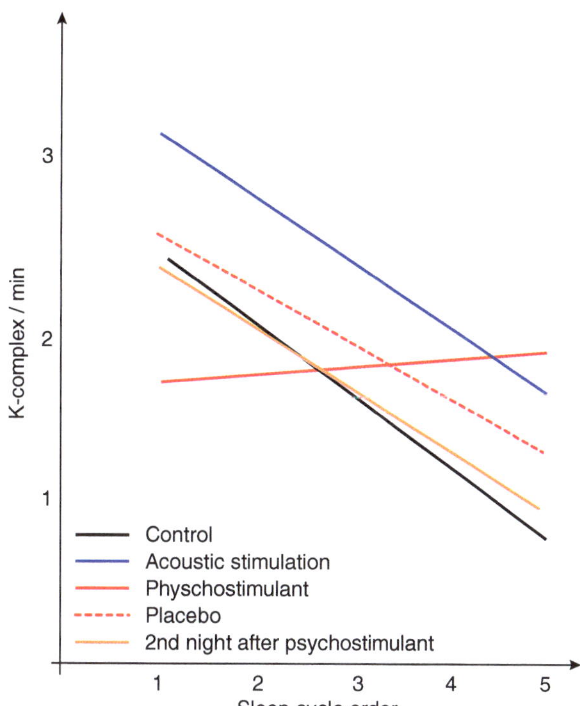

Fig. 4.2 Regression lines for average K-complex densities as functions of sleep cycle order and different treatment conditions in eight healthy adult subjects. K-complex values have shown decreasing tendencies from evening to morning in all the studied conditions except under influence of psychostimulant (parallel with the missing of the regular slow wave decay). The highest value of K-complexes was observed under acoustic stimulation (Acoustic stimulation consisted of pseudo-random 200 Hz tone clicks with durations of 10 ms and an average inter-stimulus interval of 40 s) (Modified from Halász et al. (1985))

deeper one) and the A type ones (preceded by deeper one and followed by a more superficial one) were compared, a significant difference was found in the number of K-complexes. "A" type contained more K-complexes than the "D" type. K-complex number proved to be in positive correlation with the deepness of the actual cycle. K-complex frequency decreased across the night sleep from evening to morning, according the serial order of cycles, parallel with the well-known course of the S-process of Borbély. Under the influence of psychostimulant drug, this effect ceased together with the regular course of slow wave decay (Fig. 4.2). Unlike types D, a shift in the frequency of K-complexes was very characteristic for type A stage 2.

Continuously or periodically applied random acoustic stimulation increased the number of all registered K-complexes but simultaneously decreased the number of the spontaneous ones. This change is similar to the behavior of PAT, as seen before in Chap. 3. An analysis of the K-complex number distribution during stimulation nights showed that the increase of K-complexes could mainly be attributed to an elevation of the number of type A stage 2 periods. Comparing the "A" and "D" slopes of sleep cycles, K-complex rates along proved to change more on "A" slopes and less on "D" slopes (Fig. 4.3).

Also, these studies showed again very clearly the double nature of K-complex: behaving like an arousal pattern that can be triggered by acoustic stimuli and at the same time changing parallel with the delta activity that means dampening along the sleep process, and their rate is related to the depth of the actual sleep cycle. So we can say that K-complex formation is partially associated with phasic sensorial inputs and partially with the degree of sleep synchronization, for example, depth of sleep.

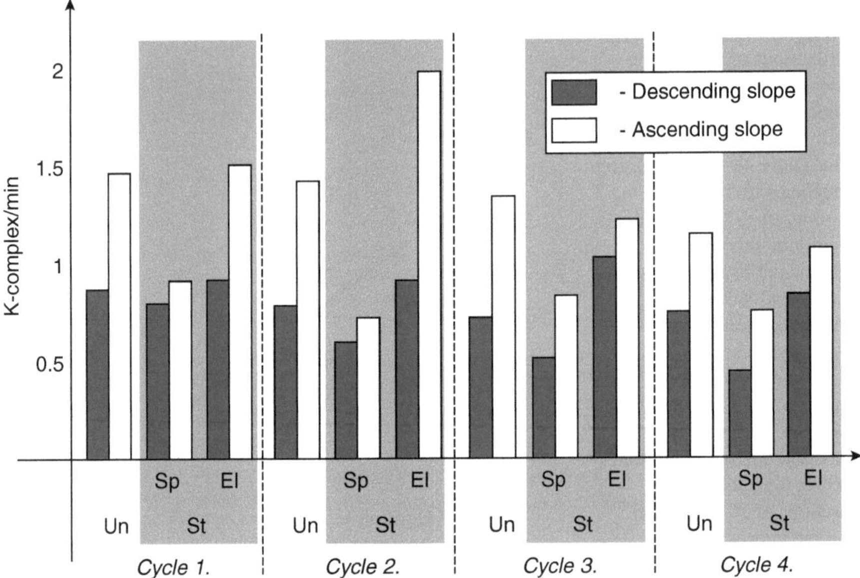

Fig. 4.3 Average number of K-complexes/min in spontaneous and stimulated stage 2 sleeps during descending and ascending slopes of a healthy young volunteer. *Un* undisturbed night, *St* stimulated night, *Sp* spontaneous K-complexes, *El* elicited K-complexes

Furthermore, these data show that K-complex is an indicator of the clash between antagonistic sleep and wake promoting forces with the result of maintaining (protection) of sleep.

The distribution of K-complexes in different parts and orders of cycles as well as the differences in their elicitability associated to dynamic phasic moments of NREM sleep throws light again to the basic instability of sleep. It unmasks the ever changing balance of sleep- and wake-promoting forces shaping the course of sleep.

Since K-complex formation is a part of slow wave activity during sleep its behavior nicely shows that the slow wave cloud during our NREM sleep is partially reactive. Later, we will see that this applies also for CAP A_1 sequences representing a much higher level beyond the K-complexes, behaving similarly to reactive sleep slow waves.

4.2.3 How Does CAP Behave Across the Cycles and During the Different Slopes of Cycles?

On the "D" slopes and especially during the first cycles, CAP A phases are less frequent than on the "A" slopes. Furthermore, the D slopes of the first cycles are characterized by the preponderance of the synchronization type, sleep-like answers (A_1 type), associated with mild autonomic perturbations (Halász 1982; Halász et al. 1985; Terzano and Parrino 2000). During the "A" slopes, phasic events and micro-arousals are more frequent, and both the EEG morphology and the associated autonomic changes are more similar to the conventional desynchronization type arousal (A_2 and A_3 type). Expressing this in terms of CAP A subtypes, the distribution of the different phase A subtypes proved to be different across the sleep cycles. Subtype A_1 occurs most frequently in the first cycles of sleep and during "D" slopes, while subtypes A_2 and A_3 are more frequent during later part of sleep across A slopes of the cycles (Parrino et al. 2012) (Fig. 4.4).

The distribution of CAP sequences across different slopes of the sleep cycles is very much similar to the distribution showed for PATs and K-complexes. The distribution of CAP A_1 (and less the A_2) phase shows similar distribution to that of PATs, whereas the distribution of A_1 (less A_2) is similar to the distribution of K-complexes. This is not surprising hence A_1 and A_2 contains K-complexes, and the second part of A_2 and mainly A_3 is identical with PATs since CAP is a global framework and overlaps with the previously described isolated phasic events and micro-arousals.

4.3 Tonic and Phasic Regulation in NREM Sleep: Relationship of the Macro- and Microstructure of Sleep Cycles

Tonic and phasic regulation works simultaneously. During "D" slope the sleep-promoting system probably has a gradually increasing predominance inhibiting the wake-promoting cell groups. Accordingly, the phasic activation is rare and results in antiarousal (sleep-promoting) response. In contrast, during slope "A" the wake-promoting system becomes gradually more dominant (Fig. 2.1). Consequently, the phasic input is more frequent and becomes more variable (antiarousal responses are present together with desynchronizational responses). Some of the micro-arousals become effective promoting state transitions as well. Therefore, during the "D" slope, the phasic input serves the avalanching sleep process. During the "A" slope, however antiarousal answers (at least in the form of K-complexes) are more frequent: reflecting the progress of the arousal process, leading to REM sleep. So the phasic input, due to the changing tonic influence, has state-dependently different effect on the "D" and "A" slopes of sleep cycles.

The tonic regulation has endogenous neuromodulatory, chemical origin, serving the biological clock and showing little flexibility to external influences. It may have NREM-promoting (during the D slope) or REM-promoting (during the A slope) nature.

On the contrary – according to our concept – phasic regulation is more related to those neural mechanisms that serve the sensory connection with the environment, characterized by faster synaptic procedures. It is more variable and more dependent on external factors.

So the "D" slopes of the cycles are more determined by the factors of the internal biochemical clock, while the "A" slopes, for example, the second part of the cycles, depend more on the connection with the external world.

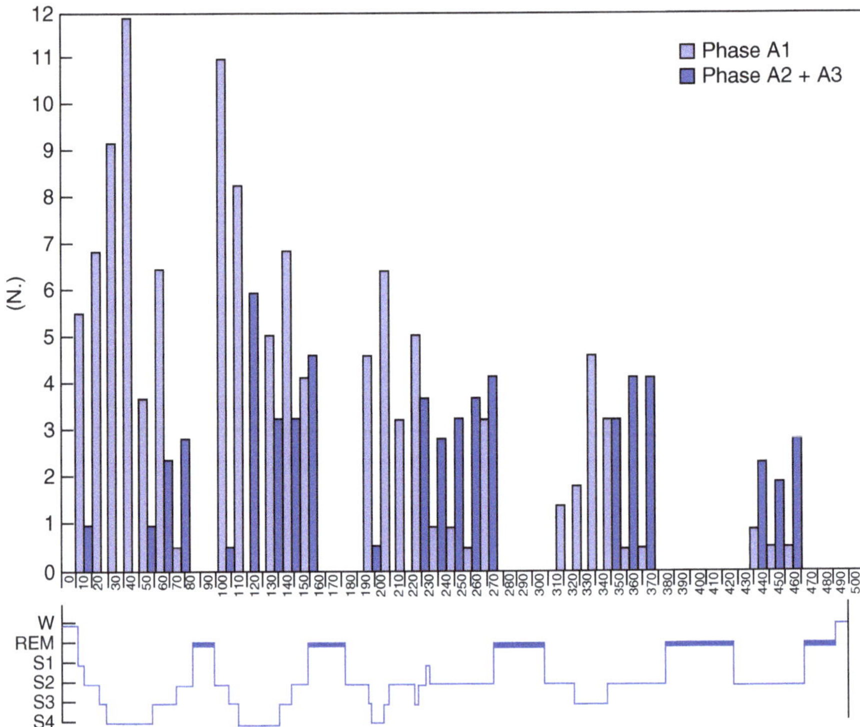

Fig. 4.4 A differential distribution of A_1 versus A_2/A_3 events over the course of sleep. Distribution of A_1 phases shows a gradual decay during each descending slope of cycles and a global decrease from evening to morning, according to the slow wave homeostatic decay. Contrasting the distribution of A_{2-3} phases have shown a peak during the ascending slopes of each cycle near before REM sleep, moving together more with the arousal process instead of the homeostatic sleep process (Based on Terzano and Parrino 2000)

Recently, we have recognized several examples of the two types of regulation working simultaneously and together. One of them provides a central (usually paired reciprocal) level setting tonic effect, while the other works following the reflex principle ensuring fast answers to the stimuli of the external world. The central regulation has a level setting tonic influence determining the state, while the other reflex-like regulations provide a flexible reactivity for the system and promote the state dynamic.[1] The two systems interact. Dynamic microstructural oscillations governed by input-dependent phasic regulation (CAP system) modulate the larger scale chemically driven oscillation (sleep cyclicity), and in the same time their elicitability and reaction type (A_1 or A_{2-3}) are state dependently determined by the larger scale shifts of the tonic regulation. The same sensory stimulus may have sleep-promoting effect on the "D"

[1] Another example of similar cooperation between the tonic and phasic systems in a certain physiological functions is seen in the relationship of brain stem (descending) influence of muscular tone regulation versus phasic activity realized in spinal reflexes.

slope and wake/REM-promoting effect on the "A" slope. Conversely, phasic activity shapes the course of tonic states providing them flexibility, promoting the state shifts.

The simultaneous tonic and phasic regulation of sleep ensures a flexible connection between the sleeper and the surrounding world and ensures the reversibility and self-regulatory nature of sleep (that differentiates sleep from comatose state).

4.4 The Nature of State Transitions and Defense Mechanisms Against Transitions (Providing Stability by Permitting Lability)

As we have shown in the previous chapters, sleep has several dynamic aspects. The sleep process contains cycles and each cycle (at least the first three to four) are built up by a "D" and an "A" slope which contains (according to the Rechtschaffen-Kales rules) sleep stages differing according to their situation on the "D" or "A" slopes of a given cycle. Stage borders are artificial, and sleep stages do not represent homogenous conditions, as was nicely shown by Merica and Fortune (1997). The stages reflect several signs of continuous deepening or lightening tendencies along the slopes of the cycles. Therefore, the transitions from one to another sleep stage seem to be stepwise: the onset of a stage bears the features of the previous one and its last part shows those of the next stage. However, under the seemingly smooth surface of sleep stages (which are merely artificially constructed conditions), there are continuous fluctuations in the form of phasic events systematized by the Parma school as a system of cyclic alternating pattern (CAP), moved by the actual balance of antagonistic sleep-promoting and arousal forces.

Contrasting to these small fluctuations, the switch from waking to sleep state and the turn from deep sleep at the through of a cycle to the "A" slope, furthermore switches to REM sleep and from REM to NREM, are larger events of the sleep/wake continuum.

As described in Chap. 1 both the wake- and sleep-promoting neuronal assemblies in the hypothalamus, like the REM-on and REM-off networks in the brain stem appear to be mutually inhibitory. Saper et al. (2001) proposed that this mutually antagonistic relationship would be similar to that seen with a so-called flip-flop electrical switch. This kind of switches ensures rapid and complete state transitions, since when either side of the antagonistic populations gains a small advantage, the inhibitory action of the other will rapidly lose its strength causing collapse of activity and an avalanche-like dominance of the other.

This kind of switch would ensure the after-switch stability both of waking or sleep and REM or NREM state. Saper et al. (2010) have arguments for the biological advantage of state stabilization saying that it may prevent an individual from falling asleep during boring activity or waking up during the night with every small sound of the house. However, in the reality we frequently fall asleep reading a not much interesting book, or wake up to noises with different significance. Sleep is differentiated from coma by arousability, which is a flexible possibility to wake up if the arousal stimulus has enough significance. Let us take the example of a prey animal. This kind of continuously tilting bilabile state ensures for the animal organism two seemingly controversial requirements: to maintain sleep separated from the environment and preserve floating attention to monitor potentially dangerous environmental events.

We have seen in the previous chapters that NREM sleep is an ever oscillating flexible state that contributes to preserve continuous contact with the environment and keeps the states in an ever changeable unsteady condition. This kind of regulation seems not realized by "bistable" states. Better to say that this regulation provides stability by permitting lability. The tilting microstates with alternating arousal or sleep-promoting oscillations ensure a higher level of stability and at the same time provide continuous possibilities for changes due to environmental demands. Therefore, in between the more deterministic state switches at the main macrostate steps, like wake to NREM and NREM to REM during a sleep cycle, the NREM states seem to behave not according to the rules of bistability predicted by the Saper model.

4.5 Promoting Forces Behind (Alternating Chemical Influences and Faster Phasic Reciprocal Sleep- and Wake-Promoting Inputs)

One of the factors determining reactivity of the organism to stimulation is the state in which the input arrives. We may speak about "state-dependent" reactivity. It is different in REM and NREM sleep and within NREM according to the "D" and "A" slope of sleep cycles (Halász 1982). A further question is the following: What are the forces underlying reciprocal tendencies during "D" and "A" slopes of the cycles and underlying shorter scale similar reciprocal tendencies reflected in phasic events and micro-arousals? Both the structure of cycles and nature of synchronizational and desynchronizational micro-arousals or in other words A_1 versus A_{2-3} CAP types may be designated as either "sleep promoting" or "wake promoting."

Let us concentrate first to the alternating tendencies observed during sleep cycles. The alternation of NREM and REM sleep was formulated by McCarley and Hobson as early as 1975 in a "reciprocal interaction" model at the cellular level between cholinergic and aminergic neuronal populations located in the brain stem (McCarley and Hobson 1975). The model was based on the contemporary findings showing firing aminergic REM-off neurons located to the locus coeruleus and dorsal raphe nucleus with an activity in waking, gradual decrease of firing during NREM, and a firing of REM-on cholinergic neuron population in the mesopontine (laterodorsalis tegmental and pedunculopontine tegmental nucleus) region during REM. A mathematical model using the Lotka-Volterra equation (originally worked out for the predator/prey interrelationship) seemed to be satisfactory explaining the detected firing relations in the aminergic and cholinergic brain stem nuclei and also to explain the state alternation on the EEG and behavioral level. Later the model was revised and adapted according to new neuromodulatory data and incorporating the influence of external noise by McCarley and Massaquoi (1992).

However, the relevance of the McCarley-Hobson model has been challenged by studies showing that lesions damaging the cholinergic and aminergic neuron populations had little effect on REM sleep (Webster and Jones 1988; Shouse and Siegel 1992; Lu et al. 2006; Blanco-Centurion et al. 2007). These data suggest only a modulator role for the cholinergic and aminergic systems in NREM/REM sleep switch. Subsequent studies revealed the key role of sublaterodorsal, pre-coeruleus and medial parabrachial nuclei in the mesopontin region. This model predicts a gradual transition to prepare the start of the REM period which may provide explanation for the changes experienced on the A slope of the NREM cycle. Accordingly, the disinhibition of the cholinergic nuclei starts at the turning point of the cycle, and the rising cholinergic activity brings the REM period closer.

Therefore, the "A" slope of the NREM cycle belongs more to the preparation phase to REM and arousal (ergotroph) activity, and the "D" slope can be considered as a slow wave sleep-promoting endophylactic process.

We have seen that each sleep stage both during the "D" and "A" slopes of the cycles is shaped by microstructural oscillations consistent with the CAP phenomena with alternative sleep- and wake-promoting micro-changes. Therefore, it seems logical to conclude that the microstructural oscillations are somehow related to sleep- and wake-promoting subcortical hypothalamic networks described by von Economo and later by the Boston (Saper and coworkers) and Lyon (Lupi and coworkers) research groups.

We may explain the wake-promoting phasic influences by periodic short-term (micro) arousals fuelled by internal and/or external sleep-perturbing impulses transitorily breaking through VLPO exerted inhibition of the RAS system. There are several proofs for the existence of such events even in neuroimaging studies (Riedner et al. 2011; Kaufmann et al. 2006; Dang-Vu et al. 2008).

The mechanism of sleep-promoting (antiarousal) phasic events (K-complex, A_1 phase) seems to be more complicated. The question is how phasic activation of the VLPO system is transferred to cortical slow wave group or K-complex? This transformation from arousing to sleep-promoting effect of a sleep-disturbing input seems to be clearly connected with the overwhelming domination of the burst-firing working mode of the thalamocortical system, after the shutting down of RAS system. The wake-promoting ascending arousal systems and the sleep-promoting VLPO system are in a reciprocal antagonistic relationship (Saper et al. 2001). When VLPO systems exert GABA- and galanin-ergic inhibitory influence and keep firing in a certain rate during sleep the inhibitory influence of the arousal system dissipates. When a phasic arousal influence is arriving, the VLPO system transitorily becomes weakly inhibited, but if the arousal influence stops, the VLPO system becomes liberated again fuelling the back inhibition of the arousal system, producing (quasi rebound) sleep phenomena by switching on the thalamocortical system (CAP A_1 phase). This holds especially true when on the "D" slopes of the first cycles the VLPO system is prevailing and the arousal system is weak. During the third part of sleep on the "A" slopes of the cycles, the situation is different: The VLPO system is probable less active in inhibiting the arousal system; therefore, the arousal impulses easily activate the cortex and the phasic arousals achieve more prominent arousals in the form of CAP A_2 and A_3 responses driving the sleeper toward more superficial vigilance states.

We will see in the next chapter how the so-called homeostatic regulation modulates sleep. The homeostatic pressure is the force which determines the course of the D slope. It keeps high the firing rate of VLPO at the beginning of D slopes, inhibiting the wake-promoting system and therefore liberating the burst-firing thalamic machine of slow wave sleep, entraining sensory stimuli as fuel for slow wave sleep by the instant homeostatic delta injections of CAP A_1 phases.

When the homeostatic pressure gradually decreases along the course of the "D" slope, this gives way to the continuously incoming sensory impulses, which start to bring the sleeper along the "A" slope to REM sleep.

References

ASDA Report. EEG arousals: scoring, rules and examples. Sleep. 1992;15(2):173–84.

Blake H, Gerard RW. Brain potentials during sleep. Am J Physiol. 1937;119:692–703.

Blanco-Centurion C, Gerashchenko D, Shiromani PJ. Effects of saporin-induced lesions of three arousal populations on daily levels of sleep and wake. J Neurosci. 2007;27(51):14041–8.

Brezinová V. Sleep cycle content and sleep cycle duration. Electroencephalogr Clin Neurophysiol. 1974;36(3):275–82.

Caille EJ, Bassano JL. Tonic structure and model of sleep distortion. Rev Electroencephalogr Neurophysiol Clin. 1976;6(2):330–4.

Dang-Vu TT, Schabus M, Desseilles M, Albouy G, Boly M, Darsaud A, et al. Spontaneous neural activity during human slow wave sleep. Proc Natl Acad Sci USA. 2008;105(39):15160–5.

Halász P. The role of the non-specific sensory activation in sleep regulation and in the pathomechanism of generalized epilepsy with generalized spike-wave discharge. Doctoral thesis, Budapest; 1982.

Halász P, Pál I, Rajna P. K-complex formation of the EEG in sleep. A survey and new examinations. Acta Physiol Hung. 1985;65(1):3–35.

Kaufmann C, Wehrle R, Wetter TC, Holsboer F, Auer DP, Pollmächer T, et al. Brain activation and hypothalamic functional connectivity during human non-rapid eye movement sleep: an EEG/fMRI study. Brain. 2006;129(Pt 3):655–67.

Lu J, Sherman D, Devor M, Saper CB. A putative flip-flop switch for control of REM sleep. Nature. 2006;441:589–94.

McCarley RW, Hobson JA. Neuronal excitability modulation over the sleep cycle: a structural and mathematical model. Science. 1975;189(4196):58–60.

McCarley RW, Massaquoi SG. Neurobiological structure of the revised limit cycle reciprocal interaction model of REM cycle control. J Sleep Res. 1992;1(2):132–7.

Merica H, Fortune RD. A neuronal transition probability model for the evolution of power in the sigma and delta frequency bands of sleep EEG. Physiol Behav. 1997;62(3):585–9.

Parrino L, Ferri R, Bruni O, Terzano MG. Cyclic alternating pattern (CAP): the marker of sleep instability. Sleep Med Rev. 2012;16:27–45.

Riedner BA, Hulse BK, Murphy MJ, Ferrarelli F, Tononi G. Temporal dynamics of cortical sources underlying spontaneous and peripherally evoked slow waves. Prog Brain Res. 2011;193:201–18.

Saper CB, Chou TC, Scammell TE. The sleep switch: hypothalamic control of sleep and wakefulness. Trends Neurosci. 2001;24(12):726–31.

Saper CB, Fuller PM, Pedersen NP, Lu J, Scammell TE. Sleep state switching. Neuron. 2010; 68(6):1023–42.

Schieber JP, Muzet A, Ferriere PJ. Phases of spontaneous transitory activation during normal sleep in humans. Arch Sci Physiol (Paris). 1971;25(4):443–65.

Shouse MN, Siegel JM. Pontine regulation of REM sleep components in cats: integrity of the pedunculopontine tegmentum (PPT) is important for phasic events but unnecessary for atonia during REM sleep. Brain Res. 1992;571:50–63.

Sinha AK, Smythe H, Zarcone VP, Barchas JD, Dement WC. Human sleep-electroencephalogram: a damped oscillatory phenomenon. J Theor Biol. 1972;35(2):387–93.

Terzano MG, Parrino L, Boselli M, Smerieri A, Spaggiari MC.CAP components and EEG synchronization in the first 3 sleep cycles.Clin Neurophysiol. 2000 Feb;111 (2):283–90.

Terzano MG, Parrino L, Rosa A, Palomba V, Smerieri A. CAP and arousals in the structural development of sleep: an integrative perspective. Sleep Med. 2002;3(3):221–9.

Webster HH, Jones BE. Neurotoxic lesions of the dorsolateral pontomesencephalic tegmentum-cholinergic cell area in the cat. II. Effects upon sleep-waking states. Brain Res. 1988; 458:285–302.

Williams RL, Agnew HW, Webb WB. Sleep patterns in young adults: an EEG study. Electroencephalogr Clin Neurophysiol. 1964;17:376–81.

Chapter 5
Changing Views of NREM Sleep Homeostatic Regulation

Abstract The homeostatic sleep regulation idea underwent important development. Sleep homeostasis was first connected to the duration of the preceding awake time. Due to increasing innovative research in this field with convincing evidences on local sleep regulation, it seems that beyond the length of waking time, use-dependent afferent stimulation and synaptic upscaling (learning) are the main factors regulating the NREM sleep slow-wave activity (SWA). Further achievement of the same research line was to obtain evidences that plastic modulation of local slow-wave power during NREM sleep is closely related to the recreation of cognitive functions in the cortex, mainly in the frontal lobes. Slow-wave homeostasis and use-dependent plasticity are probably two sides of the same coin representing the biological function of slow-wave sleep.

Keywords Sleep homeostasis • Use-dependent plasticity • Local sleep • Slow-wave activity • Delta activity

> Sleep is the price we have to pay for plasticity (Tononi and Cirelli 2003).

5.1 Discovery and First Views on Homeostatic Regulation

Homeostatic regulation in general is the control and maintenance of a stable and constant condition, achieved usually by negative feedback mechanism. One of the most powerful controls of the NREM sleep process is the homeostatic regulation expressed by the slow-wave components of sleep EEG. It is a common experience that sleep loss leads to increased sleep propensity and sleep after sleep deprivation lasts longer. Sleep loss is associated with impairment of cognitive functions and the subjective feeling of “sleepiness.” It is questionable if cognitive impairment would

P. Halász, R. Bódizs, *Dynamic Structure of NREM Sleep*,
DOI 10.1007/978-1-4471-4333-8_5,

be the consequence of lack of sleep or it is due to the sleep-inducing mechanism mobilized as a homeostatic response.

By recording a particularly high number of whole night sleep records, Webb and Agnew (1971) was the first who has unequivocally shown that the time spent in stage 4 sleep is related to the duration of presleep wakefulness. Increased time spent awake was followed by increases in stage 4 sleep time. On the other hand, it was evident that the time spent in stage 4 sleep decreases during the sleep process. Thus, a wake-dependent increase and a sleep-dependent decrease of stage 4 sleep were evidenced, suggestive for a homeostatic regulation of the SWA, which is one of the main defining features of this sleep stage. The sleep-dependent decrease of slow waves was quantified and characterized by Feinberg et al. (1978) using the period amplitude analysis measures. This investigation is also a forerunner of the concept of homeostatic regulation of sleep SWA.

A group of sleep researchers and mathematicians in Zurich led by Alexander A. Borbély has clearly proven in 1980s that the intensity of NREM sleep as measured by the power spectra of SWA depends on the duration of the previous awake state (Borbély et al. 1981). In nap studies, a gradual increase was shown in the SWA in consecutive naps during the day (Dijk et al. 1987a). This increase in slow waves of naps was the mirror image of the exponential decline of SWA during sleep, e.g., sleep propensity increases exponentially during the waking time following a slow saturating function, while during sleep, it decays exponentially (Borbély 1982; Borbély et al. 1989), reflecting the homeostatic dynamism (Dijk et al. 1987b; Fig. 5.1).

If a human being is sleep-deprived, the very next sleep will ensure the supplementation of the lost sleep. The substrate of this homeostasis was called by Borbély as "Process S," and the power spectrum of the EEG slow-wave range 0.75–4.5 Hz proved to be a good quantitative measure of it. It was shown that the patterns of sleep EEG spectral alterations induced by sleep deprivation are characterized by increased SWA (and decreased spindling). This pattern is mirrored during successive sleep cycles of undisturbed sleep. This suggests that the dissipation of sleep need during the night sleep can reliably be measured by assessing SWA (Borbély et al. 1981). Moreover, because EEG slow-wave activity is the main defining feature of slow-wave sleep, the wake-dependent exponential increase and sleep-dependent exponential decline of SWA are reliably reflected in the minutes of slow-wave sleep (Knowles et al. 1986).

The exponential increase of sleep propensity during sleep deprivation was shown to be followed by an increase in the 0.5–2-Hz slow-wave incidence during recovery sleep. These slow waves of recovery sleep were of increased frequency when compared to the ones of baseline sleep (Bersagliere and Achermann 2010). The decline of delta activity in sleep was proved to be coupled with decreased incidence of high amplitude slow waves, a decreased slope of the individual slow waves, and an increased number of multipeak waves (Riedner et al. 2007).

McCarley (2007) have shown that adenosine acting in the basal forebrain is a putative mediator of homeostatic control. Increased need for sleep was accompanied by increased release of adenosine. In other experiments, injection of adenosine or an adenosine A1 receptor agonist into the rat basal forebrain or cat VLPO promoted sleep by inhibiting wake-promoting regions (Scammell et al. 2000; Strecker et al. 2000). The relationship between adenosine and sleep homeostasis is evident

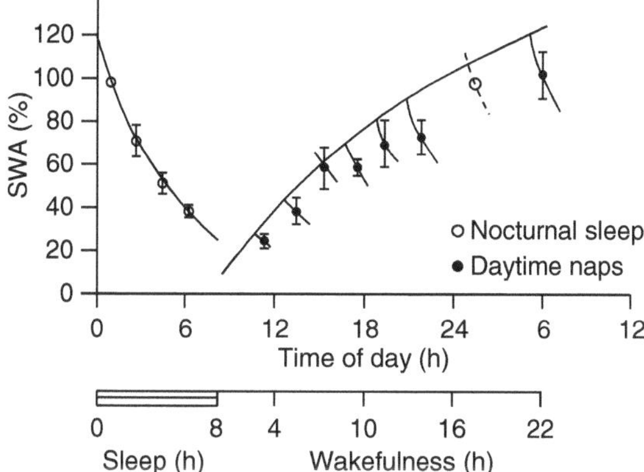

Fig. 5.1 Homeostatic regulation of sleep slow-wave EEG activity (0.75–4.5 Hz). Power of sleep slow-wave activity follows an exponentially decreasing trend during nocturnal sleep. On the other hand, daytime nap studies reveal the slowly saturating exponential function describing the increase of slow-wave activity in the absence of sleep. *SWA* slow-wave activity (Dijk et al. 1987a)

from the studies investigating the effects of naturally occurring functional polymorphisms of the gene coding the adenosine deaminase (ADA) enzyme in humans. The heterozygous G/A variant of the 22 G > A polymorphism of the gene encoding ADA results in reduced adenosine breakdown and consequent increases in the nightly amounts of slow-wave sleep and slow-wave EEG activity during both NREM and REM sleep (Rétey et al. 2005). Although similar results were revealed in studies on rats, it is premature to conclude that adenosine is the only and main neurochemical substrate of sleep homeostasis. However, as adenosine is a result of the breakdown of ATP, it is evident that the brain's energy balance is involved in the homeostatic regulation of the sleep process. Other neurochemical factors involved in sleep homeostasis are the somnogenic cytokines (interleukin-1 beta [IL1-β] and tumor necrosis factor alpha [TNF-α]) as well as the molecules involved in long-term potentiation (LTP), like the brain-derived neurotrophic factor (BDNF).

The main aspect was the amount of prior wakefulness, considered by Wilse B. Webb and Alexander A. Borbély. This finding was coherent with the hypnotoxin theory of Henri Pieron (1913), suggesting that a wake-dependent release of some somnogenic chemical agent accounts for the accumulation of sleep need.

5.2 Use-Dependent Homeostatic Regulation and Local Plasticity: Sleep-Dependent Improvement of Learning and Plasticity

In the last 10–15 years, local aspects of homeostatic regulation have received more and more attention: the frontal preponderance of sleep slow-wave activity and frontal dominance of the recovery increase after sleep deprivation and dominant hemisphere preponderance (Achermann et al. 2001) were emphasized in several studies (Finelli et al. 2001; Cajochen et al. 1999; Marzano et al. 2010). This has been interpreted as evidence for the role of slow-wave sleep in human frontal cognitive functions (Horne 1993). Later, when local aspects of sleep regulation were increasingly evident, the concept of wake dependency was slowly completed with the notion of use dependency, which is in fact based on the amount of afferent inputs to a certain neural network. That is, equal amounts of wakefulness were shown to be followed by topographically different increases in SWA. Differences were attributed to the differences of use of the neural systems involved in processing the afferent inputs during wakefulness. Use dependency is in part a heritage of the infinite search for the hypnotoxin proposed by Pieron or the S-factor of Pappenheimer et al. (1975). The attempts to characterize this chemical agent have lead to the discovery of the strong link between sleep and the immune system. The slow-wave sleep-inducing effects of some somnogenic cytokines, like the TNF-α, were shown to be local. Local applications lead to increases in local sleep SWA (Yoshida et al. 2004).

Besides the frontal lobes involved in homeostatic regulation related to certain cognitive functions, other different regions producing use-dependent increase of SWA have been registered after specific use. Extensive sensory stimulation of one hand before sleep led to an increase of sleep delta power in the opposite hemisphere over the somatosensory arm area (Kattler et al. 1994). An opposite intervention, immobilization of the arm, caused a local reduction of delta power in the same localization (Huber et al. 2006). Similar results were found in rats after cutting their whiskers on one side and analyzing the changes in hemispheric asymmetry in their sleep SWA (Vyazovskiy et al. 2000).

Further aspects of local sleep concept came from the work of Krueger and Obál (1993), who showed indirect and direct evidences indicating that cortical columns oscillate between functional states defined by changing input-output relationship tested by amplitudes of evoked responses (Rector et al. 2009). Mosaics of sleeping columns can be found while other columns are continuously awake. The longer the column is in awake-like state, the higher the probability that it will turn to sleep. The probability to find sleeping columns also depends on the amount of afferent activity or on the neuronal traffic resulting in learning. Krueger and Obál (1993) emphasized that sleep is a statistical phenomenon: e.g., a sum of the local sleep processes leads to global/behavioral sleep if there is a sufficiently high number of neural networks involved. Krueger et al. (2008) suggest that global coordination of NREM is not due to a single sleep generator, but may reflect an emergent property of loosely coupled local processes.

Local sleep regulation is a crucial point in understanding the nature of sleep homeostasis. The original two-process model and its later specifications (Borbély 1982; Borbély et al. 1989) focused on global aspects of the process only. However, it is evident from the above-cited studies that local differences are significant and robust enough for considering their involvement in the global aspect of sleep regulation or sleep need. According to this view, sleep regulation centers are "just" coordinators, providing a more or less synchronized entry of many different networks into the sleep state. Sleep-inducing centers, like the VLPO region, are not per se sleep inducers, but synchronizers of many local sleep needs. Another view is that of Zavada, who changed the original two-process model in order to cover the phenomena of local sleep intensities (Zavada et al. 2009). The Process Z would be the one which is involved in local sleep regulation. In this view, there is a global sleep need (quantified by the Process S), which is independent from the local fluctuations of sleep intensity (Process Z).

Later, learning and synaptic plasticity was shown to be related to sleep SWA, and the notion of experience dependency emerged, suggesting that it is the net change in synaptic strength that is important for sleep homeostasis (Tononi and Cirelli 2003). Experience dependency means that equal amounts of wakefulness and afferent stimulation could lead to different rates of increase in sleep SWA if different learning rates were present during wakeful afferent stimulation (Huber et al. 2004). Obviously, the amounts of wakefulness and afferent activity are important factors in determining the net synaptic changes, but they are not the only and not the main determining factors. In fact, the importance of the quality of wakefulness, the amount of new experiences, that subjects were faced during wakefulness, was shown to alter the subsequent slow-wave sleep, in a field study of Horne and Minard (1985). Subjects of the experiments of this study were unexpectedly involved in different playful activities: they had a car journey to another city, they visited a large exhibition center and a museum, and they were invited for a whole-day program in an amusement park and a zoo instead of boring paper-and-pencil tests. Although physical activity did not enhance during these programs, their effect on the subsequent slow-wave sleep was evident during the sleep laboratory examination (Horne and Minard 1985).

Stickgold and coworkers (2000) have confirmed that some kind of learning (texture discrimination task) may occur only after a night of sleep. Sleep-deprived subjects fail to improve on texture discrimination even after two recovery nights. SWA is the tool used by sleep for this process. If we demolish SWA by acoustic stimulation (Aeschbach et al. 2008), the subsequent improvement after daytime training will be lost (Fig. 5.2).

So, recent studies have shown that procedures presumably leading to local plastic changes in the cerebral cortex can lead to local changes in SWA during subsequent sleep.

The ontogenetic aspect of sleep-dependent learning came also in the focus of research. Early sensory deprivation in animal's life reduces sleep SWA (Miyamoto et al. 2003). Huber et al. (2008) used median nerve paired associative stimulation followed by transcranial magnetic stimulation (TMS) pulses to the contralateral cortical hand area. This procedure was leading to LTP or depression of cortical excitability. During subsequent sleep, SWA increased or decreased locally in subjects answering with increase or decrease of cortical response, respectively. Furthermore, during subsequent sleep, SWA increased locally after TMS had increased and decreased after TMS had decreased following paired stimulation. Changes in TMS evoked cortical EEG response and changes in SWA were localized to similar cortical regions and correlated positively. So they experienced again a tight relationship between cortical plasticity and sleep intensity.

Generalizing these results, it seems that sleep is a plastic process affected by waking experiences. On the other hand, the above experiences clearly show that slow waves, also associated with plastic changes, may be reactive elements of NREM sleep.

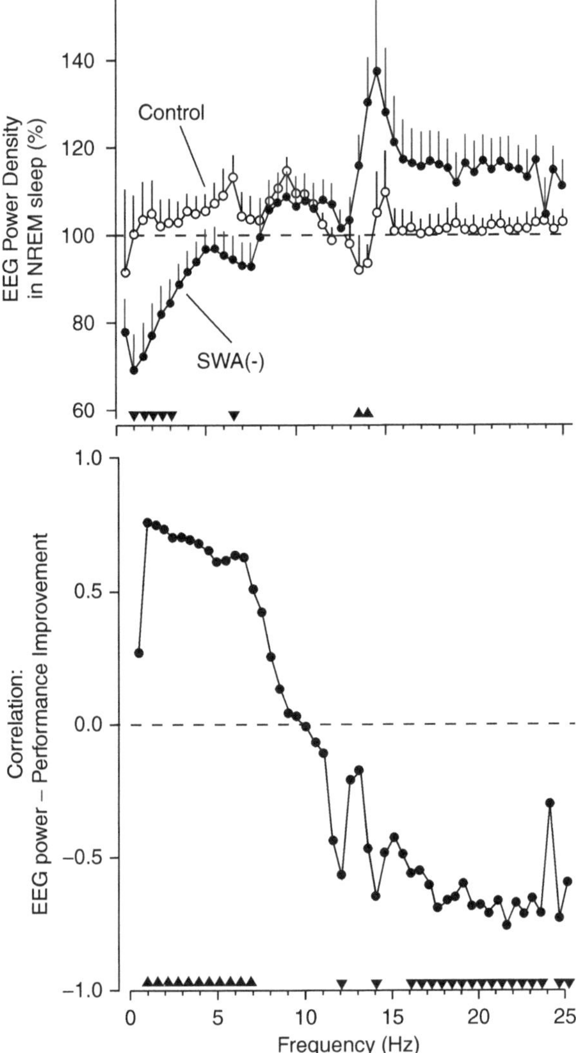

Fig. 5.2 Sleep slow-wave activity as a tool to improve overnight performance in a visual texture discrimination task. *Top*: slow-wave sleep-disruptive acoustic stimulation significantly reduced sleep slow-wave activity in the occipitoparietal region (derivation O2P4) in healthy adult volunteers. *Bottom*: power in the low EEG frequencies of O2P4 derivation correlates positively with overnight performance and improvement in a visual texture discrimination task. Filled triangles, $p < 0.05$ (Aeschbach et al. 2008)

Another interesting aspect of the interrelationship between sleep and cognitive functions is how day and night functions are interconnected. Daytime use of a function (e.g., a cognitive task) led to an increase in slow-wave sleep intensity over the cortical representation of the function. This increase is associated by an improvement of the same cognitive performance. Thus, day and night (waking and sleeping) are strongly interconnected, and sleep supports the learning process.

Therefore, there are more and more evidences supporting the view that NREM sleep, and especially slow-wave sleep homeostatic regulation, is governed by use-dependent plasticity processes. In other words, slow-wave homeostasis and

Table 5.1 The changing views of sleep homeostasis

Theoretical framework	Nature of sleep homeostasis	Key factor in sleep homeostasis	Experimental protocol	Neurochemical background (antecedents)	Key references
Two-process model	Sleep-wake-dependent	Presleep wakefulness	Constant routine	Hypnotoxin theory	Borbély et al. (1981), Borbély (1982)
Neuronal group theory	Use-dependent	Afferent stimulation	Presleep stimulation	Somnogenic cytokines	Krueger and Obál (1993), Kattler et al. (1994)
Synaptic homeostasis	Experience-dependent	Learning	Presleep learning	LTP-related neurochemical factors	Tononi and Cirelli (2003), Huber et al. (2004)

use-dependent plasticity are probably two sides of the same coin representing the biological function of slow-wave sleep. The full-blown development of this intermingled regulation is most probably a human neoformation due to the high cognitive functions and vulnerability of the frontal neocortex.

Thus, the discoveries of local sleep regulation lead sleep researchers to change their views on sleep homeostasis. These changes are summarized in Table 5.1.

References

Achermann P, Finelli LA, Borbély A. Unihemispheric enhancement of delta power in human frontal sleep EEG by prolonged wakefulness. Brain Res. 2001;913(2):220–3.

Aeschbach D, Cutler AJ, Ronda JM. A role for non-rapid-eye-movement sleep homeostasis in perceptual learning. J Neurosci. 2008;28(11):2766–72.

Bersagliere A, Achermann P. Slow oscillations in human non-rapid eye movement sleep electroencephalogram: effects of increased sleep pressure. J Sleep Res. 2010;19(1 Pt 2):228–37.

Borbély AA. A two process model of sleep regulation. Hum Neurobiol. 1982;1(3):195–204.

Borbély AA, Baumann F, Brandeis D, Strauch I, Lehmann D. Sleep deprivation: effect on sleep stages and EEG power density in man. Electroencephalogr Clin Neurophysiol. 1981;51(5): 483–95.

Borbély AA, Achermann P, Trachsel L. Tobler sleep initiation and initial sleep intensity: interactions of homeostatic and circadian mechanisms. J Biol Rhythms. 1989;4(2):149–60.

Cajochen C, Foy R, Dijk DJ. Frontal predominance of a relative increase in sleep delta and theta EEG activity after sleep loss in humans. Sleep Res Online. 1999;2(3):65–9.

Dijk DJ, Beersma DG, Daan S. EEG power density during nap sleep: reflection of an hourglass measuring the duration of prior wakefulness. J Biol Rhythms. 1987a;2(3):207–19.

Dijk DJ, Beersma DG, Daan S, Bloem GM, Van den Hoofdakker RH. Quantitative analysis of the effects of slow wave sleep deprivation during the first 3 h of sleep on subsequent EEG power density. Eur Arch Psychiatry Neurol Sci. 1987b;236(6):323–8.

Feinberg I, March JD, Fein G, Floyd TC, Walker JM, Price L. Period and amplitude analysis of 0.5-3 c/sec activity in NREM sleep of young adults. Electroencephalogr Clin Neurophysiol. 1978;44(2):202–13.

Finelli LA, Borbély AA, Achermann P. Functional topography of the human nonREM sleep electroencephalogram. Eur J Neurosci. 2001;13(12):2282–90.

Horne JA. Human sleep, sleep loss and behaviour. Implications for the prefrontal cortex and psychiatric disorder. Br J Psychiatry. 1993;162:413–9.

Horne JA, Minard A. Sleep and sleepiness following a behaviourally 'active' day. Ergonomics. 1985;28(3):567–75.

Huber R, Ghilardi MF, Massimini M, Tononi G. Local sleep and learning. Nature. 2004;430(6995):78–81.

Huber R, Ghilardi MF, Massimini M, Ferrarelli F, Riedner BA, Peterson MJ, Tononi G. Arm immobilization causes cortical plastic changes and locally decreases sleep slow wave activity. Nat Neurosci. 2006;9(9):1169–76.

Huber R, Määttä S, Esser SK, Sarasso S, Ferrarelli F, Watson A, Ferreri F, Peterson MJ, Tononi G. Measures of cortical plasticity after transcranial paired associative stimulation predict changes in electroencephalogram slow-wave activity during subsequent sleep. J Neurosci. 2008;28(31): 7911–8.

Kattler H, Dijk DJ, Borbély AA. Effect of unilateral somatosensory stimulation prior to sleep on the sleep EEG in humans. J Sleep Res. 1994;3(3):159–64.

Knowles JB, MacLean AW, Salem L, Vetere C, Coulter M. Slow-wave sleep in daytime and nocturnal sleep: an estimate of the time course of "process S". J Biol Rhythms. 1986;1(4):303–8.

Krueger JM, Obál F. A neuronal group theory of sleep function. J Sleep Res. 1993;2(2):63–9.

Krueger JM, Rector DM, Roy S, Van Dongen HP, Belenky G, Panksepp J. Sleep as a fundamental property of neuronal assemblies. Nat Rev Neurosci. 2008;9(12):910–9.

Marzano C, Ferrara M, Curcio G, De Gennaro L. The effects of sleep deprivation in humans: topographical electroencephalogram changes in non-rapid eye movement (NREM) sleep versus REM sleep. J Sleep Res. 2010;19(2):260–8.

McCarley RW. Neurobiology of REM and NREM sleep. Sleep Med. 2007;8(4):302–30.

Miyamoto H, Katagiri H, Hensch T. Experience-dependent slow-wave sleep development. Nat Neurosci. 2003;6(6):553–4.

Pappenheimer JR, Koski G, Fencl V, Karnovsky ML, Krueger J. Extraction of sleep-promoting factor S from cerebrospinal fluid and from brains of sleep-deprived animals. J Neurophysiol. 1975;38(6):1299–311.

Pieron H. Le problème physiologique du sommeil. Paris: Masson; 1913.

Rector DM, Schei JL, Van Dongen HP, Belenky G, Krueger JM. Physiological markers of local sleep. Eur J Neurosci. 2009;29(9):1771–8.

Rétey JV, Adam M, Honegger E, Khatami R, Luhmann UF, Jung HH, Berger W, Landolt HP. A functional genetic variation of adenosine deaminase affects the duration and intensity of deep sleep in humans. Proc Natl Acad Sci USA. 2005;102(43):15676–81.

Riedner BA, Vyazovskiy VV, Huber R, Massimini M, Esser S, Murphy M, Tononi G. Sleep homeostasis and cortical synchronization: III. A high-density EEG study of sleep slow waves in humans. Sleep. 2007;30(12):1643–57.

Scammell TE, Estabrooke IV, McCarthy MT, Chemelli RM, Yanagisawa M, Miller MS, Saper CB. Hypothalamic arousal regions are activated during modafinil-induced wakefulness. J Neurosci. 2000;20(22):8620–8.

Stickgold R, James L, Hobson JA. Visual discrimination learning requires sleep after training. Nat Neurosci. 2000;3(12):1237–8.

Strecker RE, Morairty S, Thakkar MM, Porkka-Heiskanen T, Basheer R, Dauphin LJ, Rainnie DG, Portas CM, Greene RW, McCarley RW. Adenosinergic modulation of basal forebrain and preoptic/anterior hypothalamic neuronal activity in the control of behavioral state. Behav Brain Res. 2000;115(2):183–204.

Tononi G, Cirelli C. Sleep and synaptic homeostasis: a hypothesis. Brain Res Bull. 2003 Dec 15;62(2):143–50.

Vyazovskiy V, Borbély AA, Tobler I. Unilateral vibrissae stimulation during waking induces interhemispheric EEG asymmetry during subsequent sleep in the rat. J Sleep Res. 2000; 9(4):367–71.

Webb WB, Agnew Jr HW. Stage 4 sleep: influence of time course variables. Science. 1971; 174(4016):1354–6.

Yoshida H, Peterfi Z, García-García F, Kirkpatrick R, Yasuda T, Krueger JM. State-specific asymmetries in EEG slow wave activity induced by local application of TNFalpha. Brain Res. 2004; 1009(1–2):129–36.

Zavada A, Strijkstra AM, Boerema AS, Daan S, Beersma DG. Evidence for differential human slow-wave activity regulation across the brain. J Sleep Res. 2009;18(1):3–10.

Chapter 6
Homeostatic Features of the CAP System and the Physiological Mechanism of Reactive Slow Wave Activity

Abstract The significance of reactive (input-related) slow wave responses has remained hidden for a long time. The recognition of this phenomenon started with the discovery of K-complexes and continued with the description of the A_1 phase of cyclic alternating pattern (CAP); however, their biological role and importance is not yet fully understood. We propose here that these phenomena represent an instant (short-term) slow wave homeostasis to ensure the necessary amount of slow waves of a night. We propose that this instant slow wave homeostatic regulation is a supplementary regulation tool of slow wave economy as a response to any perturbation threatening the realization of the necessary slow wave amount especially during the first cycles, when homeostatic sleep pressure is high. There we have the link between the arousal-related microstructural sleep regulation and the sleep slow wave homeostasis, believed hitherto to be different issues.

Keywords CAP • Instant slow wave homeostasis • Delta injection • Sensory evoked responses • K-complex • Boosting of slow waves

> While during wakefulness corticothalamic circuits react to a direct perturbation with a complex pattern of activation, during NREM sleep the only possible response is a stereotypical slow wave, that can be local or global, depending on stimulation intensity (Massimini).

6.1 CAP System and Sleep Homeostasis: Is CAP an Instant Homeostatic Tool?

The discovery of CAP in the mid-eighties by the Parma sleep research group (Terzano et al. 1985) and later studies have shown that this system is a very sensitive marker of NREM instability (Parrino et al. 2012). The distinction among different types of A phases (Terzano et al. 2000) and the contrasting functional behavior of

DOI 10.1007/978-1-4471-4333-8_6,

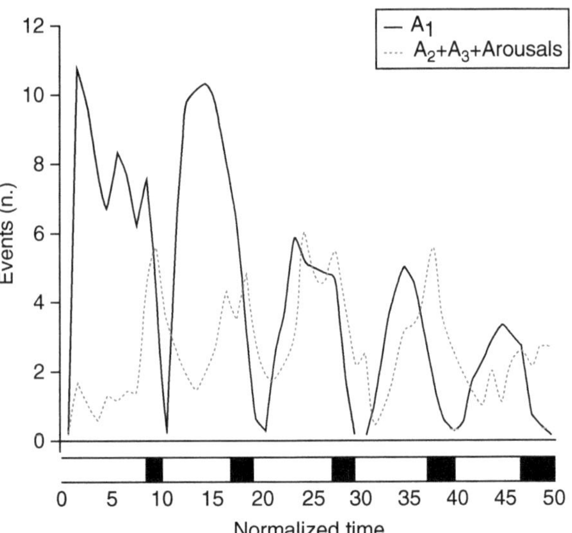

Fig. 6.1 Involvement of CAP A subtypes in the homeostatic and ultradian regulation of sleep. CAP A_1 events follow the dampened sinusoid oscillation reflecting the exponential decrease in slow waves as predicted by the two process models of sleep regulation of Borbély. On the other hand, CAP A_{2-3} subtypes and arousals are peaking shortly before REM sleep (*black boxes* at the *bottom of the figure*) without showing any sleep-dependent decrease (Terzano et al. 2005)

A_1 and A_3 types set the stage for further development. The A_1 type has been identified as sleep-promoting (antiarousal) event fueling slow wave sleep and behaving similarly as spontaneous slow waves. On the contrary, A_3 has been identified as a PAT analogue, arousal-like event. These contrasting features of CAP constituents were well congruent with our earlier distinction of "synchronizational" and "desynchronizational" type microarousals. The different types of their distribution across the sleep cycles also support their different natures (Halász et al. 2004).

Terzano et al. (2005) analyzed how CAP is involved in the homeostatic and ultradian sleep dynamics. CAP A_1 rate in NREM sleep undergoes a characteristic significant exponential reduction during night sleep parallel with the homeostatic decay of slow wave activity according to the Borbély's S process (1982). This is congruent with our results on K-complexes (Halász 2005) showing similar time course across the sleep process. The same effect of the homeostatic process has been shown by Vyazovskiy et al. (2009) finding the alteration of the amplitude of slow waves, elicited by intracortical electrical stimulation during NREM sleep (see Chap. 7). Ferri et al. (2005) have shown that CAP A_1 is involved in the gradual synchronization of EEG during sleep. In contrast to this, A_{2-3} rates have shown quite different distribution with cyclic peaks before the REM periods on the ascending ("A") slopes of cycles, without any dampening during the course of sleep (Fig. 6.1). This means that input-dependent A_1 events have a part in building up slow wave sleep and follow the course of the homeostatic process. A_{2-3} events, on the contrary, belong to the wake/REM-promoting system taking part in arousal regulation of NREM sleep.

As mentioned earlier, the CAP system was considered by the Parma group as a marker of sleep instability on one hand and as a buffer system against perturbations of NREM sleep on the other. Both seem true, but the "instability" aspect is more attached to the dynamics of A_2–A_3 events, and the "buffer system" aspect is more attached to the dynamics of the A_1 events. Similarly to slopes of sleep cycles exhibiting double nature "sleep promoting" during the descending and "wake/REM promoting" during the ascending slopes, CAP constituents follow the same alternating trend (Fig. 6.1).

Furthermore, if we try to check the biological role of the CAP system, the "instability marker" role seems to be a medical consideration, emphasizing abnormality as something deviating from the physiological. We propose to look on the other side of the coin and rather examine the "buffer system" aspect. If we do so, a very interesting feature of the CAP system is coming to light. We may consider the system as a short-range homeostatic process in which the amount of slow wave activity is instantly buffered and sleep continuity is kept preserved. We may comprehend the CAP system as a natural "delta injection," fuelling the defense of sleep against perturbations.

With the recognition of the CAP A_1 features, we have learned that the flow of slow waves is not continuous even along the descending slope of the first cycles, but appears in input-related bouts. Depending on the degree of sleep perturbation, 30–40 % of the sleep time is spent usually in CAP A phase (Terzano et al. 2000). The CAP sequences are characterized by recurrent bouts of reactive slow activity (main frequency 0.25–2.5 Hz) represented by the A_1 phase amounting more than 60 % of all the CAP sequences. This rate is an overall sleep average. The participation of slow waves increases in deep sleep and the range of A_1 type phases among the A subtypes also increases; thus, the specific rate of reactive slow wave activity in deep sleep is much higher.

A not much emphasized aspect demonstrated by the CAP phenomenon is the input-related nature of an important part of NREM sleep slow wave activity. If a sensory stimulus is applied during B phase, an A phase event is elicitable. The type of this A phase response depends on the actual sleep pattern receiving the disturbing stimulus according to cycle, stage, and slope orientation.

So we have to realize that an important compartment of the slow wave cloud over the sleeping brain has an input-related origin linked to the CAP system. The Terzano study, connecting the A_1 type with the homeostatic process, enlightens another important aspect, namely, A_1 power reflects the actual level of the homeostatic decay of slow waves; consequently, it is cortical (metabolic) demand dependent.

The early experiments of the Parma group (Terzano et al. 1988) clearly showed that the slow wave power of A_1 and A_2 phases increased proportionally under the increasing sound pressure levels of 45–75 dB white noise; the fast frequencies' power during phase A_3 remained unchanged. So the increase of the reactive slow wave rate seems to be proportional with the degree of sleep perturbation caused by external sound (Fig. 6.2).

When sleep delta power increases after a sleep-deprived night, the CAP system reacts with a robust decrease of its rate (De Gennaro et al. 2002). The supplementation of sleep with slow waves during the night after sleep deprivation obstacles the production of CAP sequences: probably because the sleep-promoting system is being saturated.

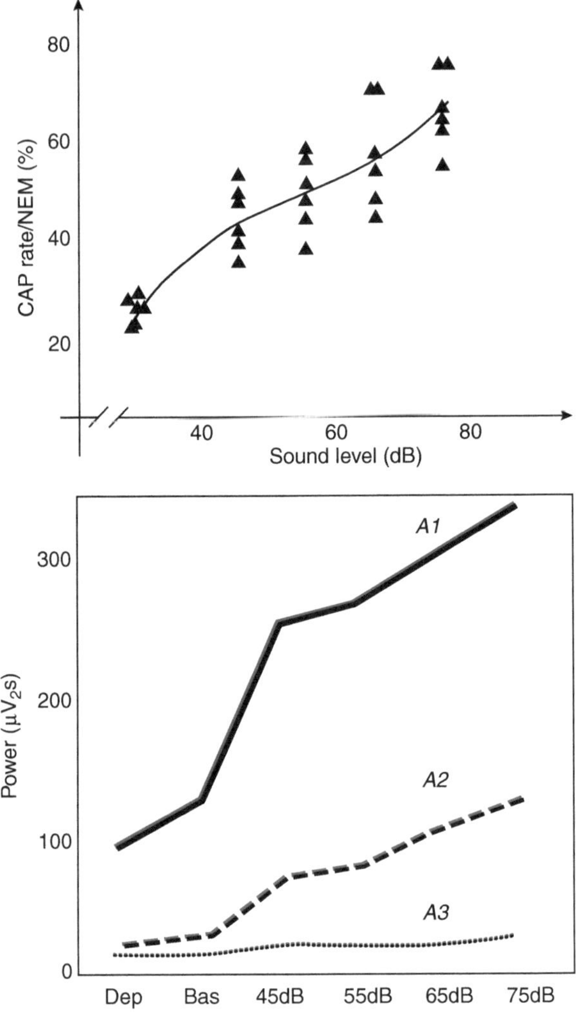

Fig. 6.2 The input and sleep pressure dependence of the cyclic alternating pattern (*CAP*). *Top*: Increasing level of sound (acoustic stimulation with white noise) lead to increases in CAP rate. *Bottom*: Power of the CAP A_1 and A_2 subtypes (containing slow waves) increases with increasing levels of sound (acoustic stimulation with white noise), while power of A_3 (containing fast EEG activity mainly) remains relatively unchanged. Sleep deprivation decreases the power of input-related (reactive) delta bouts. *Dep* deprivation, *Bas* baseline

We have obtained similar results on K-complex production under the influence of random sound stimulation (Halász 1982, page 70). During the stimulation nights, the frequency of K-complexes (spontaneous + evoked) has increased, while the number of spontaneous ones decreased (Fig. 4.3). Interestingly, when the frequency of sound stimuli was further increased (until 17/min), the elicitability of K-complexes broke down, while the slow wave power of the background activity increased (Fig. 6.3).

Here, again we are emphasizing that the input-dependent slow wave supplement within the frame of the CAP system provides an instant homeostatic function that ensures the necessary sleep protection by allowing lability (and reversibility) instead of bistability, as proposed by Saper et al. (2001).

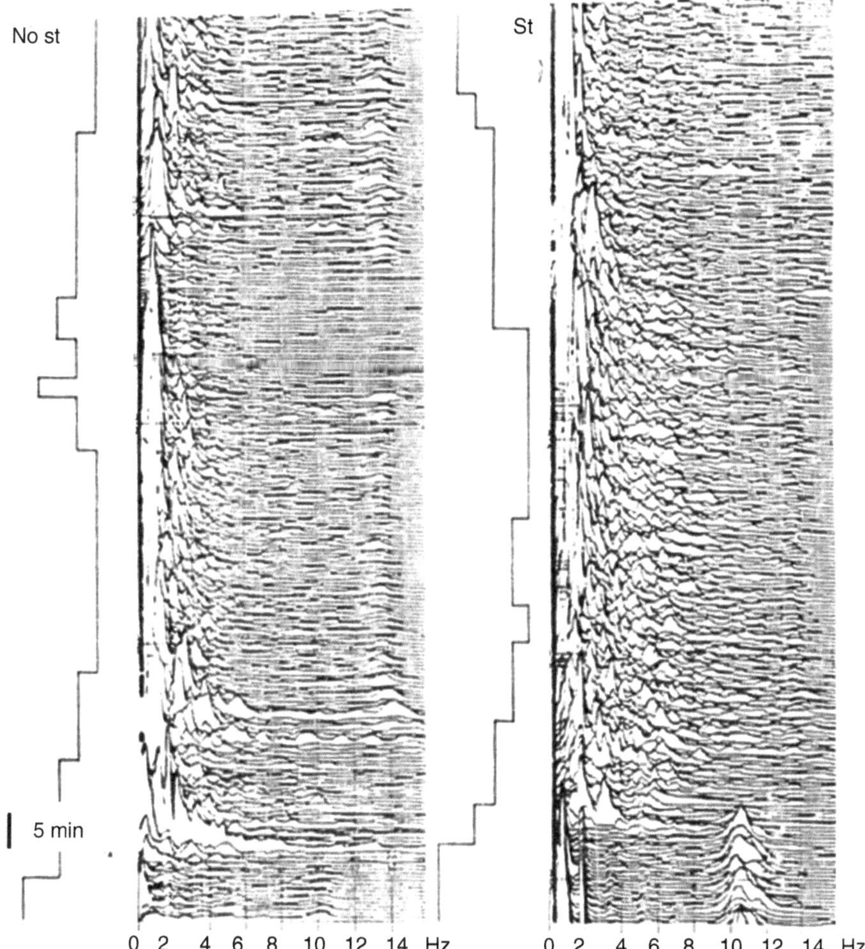

Fig. 6.3 Compressed spectral array of one cycle of NREM sleep EEG. *Left*: spontaneous sleep. *Right*: acoustic stimulation at the frequency of 17/min. Under stimulation power of slow wave activity increased. *St* stimulation, *No St* no stimulation

An important feature of the instant homeostatic role of CAP A_1 phases is their timing, which is mainly limited to the descending slopes of the first cycles during homeostatic decay of slow wave production. This time window seems to be the main possibility to get the necessary amount of slow waves: if arousing events decrease slow wave production, the CAP system provides supplementary slow waves. The supplementation of sleep with slow waves during the night after sleep deprivation obstacles the production of CAP sequences: probably because the sleep-promoting system is being saturated. Another also important feature of the instant homeostatic function is that it is targeted to the frontal lobe as was shown by Ferri (2006). Therefore, it seems that the frontal lobe is under a double homeostatic regulation, probably because this sensitive, phylogenetically new region needs slow wave activity for the recuperation of cognitive functions processed mostly here.

Presently, we do not know whether this is due to the characteristics of input factors (left dominance, e.g., might be associated with the greater metabolic demand in the cortical areas of the speech network) or to the sensitivity (amount of restitutional need) of certain cortical areas (e.g., frontal areas responsible for more sensitive human cognitive neofunctions).

We should remind the reader about the earlier touched distinction between tonic and phasic regulation. So we may assume that here, similarly to other physiological regulations, again a double, tonic, and phasic regulation works together. The traditional long-term (exhibited during the next sleep period) homeostatic regulation is a slow chemically driven process, contrasting to the short-term (instant) slow wave regulating homeostatic process which is faster and more synaptically driven. Beyond the differences, the two homeostatic processes are interrelated: the instant process is built upon the traditional one that borrows force from the strength of the homeostatically tuned thalamocortical system and is timed to the high homeostatic pressure period of the cycle. The common subject of the double homeostasis, the slow wave activity.

As it was enumerated in the previous chapters, more and more evidences support the view that NREM sleep, and especially slow wave sleep homeostatic regulation, is in the same time governed by use-dependent plasticity processes. In other words, slow wave homeostasis and use-dependent plasticity seem to be closely related, supporting the restitution of those cortical areas which have the highest vulnerability, engaged with cognitive functions.

Based on results of contemporary sleep research, the regulation of NREM sleep can be integrated in a more uniform concept as earlier. Arousal regulation, homeostatic regulation, and plastic changes of sleep all can be viewed as complex relationships around reactive use-dependent regional slow wave activity of sleep.

6.2 Mechanism of Sleep-Like Response in NREM: Relationship with the Late Slow Wave Components of Sensory Evoked Responses

The precise mechanism of input-related (reactive) slow wave responses either in the form of K-complex or CAP A_1 phase is presently unknown, but we may delineate some of the constituents of the complex mechanism underpinning this response.

To understand more, first we should discuss the transfer of information from sensory input to cortical perception during sleep compared to the waking state. Sensory systems reacting to environmental stimuli convey impulses to the thalamus and from the thalamus to the cortex. During the waking state, the complete message of the peripheral receptors reaches the sensory cortical areas: the input and output ratio of the thalamic cells is almost one (Coenen 1995). During sleep from drowsiness to deep NREM sleep, due to the decreasing output rate of the more and more hyperpolarized thalamic relay cells as spindling and later slow wave sleep develops, the transfer ratio falls down progressively to 0.3–0.4. Several human studies showed that although sensory inputs after peripheral stimulation pass the thalamic gate during all phases of NREM sleep, due to the burst-firing working mode of the thalamocortical system, the membrane potentials fluctuate both in thalamic and cortical neurons, and during the down state of cortical slow oscillation, the sensory processing becomes nonstationary and impaired (Massimini and Rosanova 2003; Rosanova and Timofeev 2005; Schabus et al. 2012).

Studies on evoked responses during the wake-sleep cycle have a long-lasting history since 1939 (Davis et al. 1939). The sensory evoked potentials (ERP) are divided to early and late components. The early waves are determined by the physical qualities (intensity, duration) of the stimulus, while the late ones are influenced by endogenous factors related to the elaboration of the sensory information. Late components are particularly sensitive to vigilance changes. The increased synchronization in slow wave sleep due to the gating by burst-firing thalamocortical neurons contributes to the enlargement of the late ERS components, taking shape of large polyphasic slow waves. The most frequently studied evoked potential in sleep is the auditory evoked potential.

In a previous chapter, we enumerated arguments for K-complex being considered analogous with the late nonspecific components of evoked potentials in NREM sleep. The recent fMRI work of Riedner et al. (2011) has confirmed this relationship.

The functional relationship of K-complex components with slow wave oscillation of NREM sleep was evidenced congruently from different aspects (Plihal et al. 1996; Amzica and Steriade 1998; Peszka and Harsh 2002; Nicholas et al. 2002; Halász 2005; Cash et al. 2009). Nowadays, it is clear that sensory stimuli "act as a trigger pulse producing a resonance in the synchronized neural assembly" of slow sleep oscillation (Coenen 1995). In a recent work, the entrainment of the slow oscillation of auditory thalamic neurons was explicitly proven by repetitive sound stimuli (Gao et al. 2009) which were able to drive the timing of cortical up-down state oscillation during slow wave sleep.

By recording intracortical field potentials by chronically implanted multielectrodes in freely moving cats, Karmos et al. (1988) have found that the middle latency positive-negative wave that followed the early components of acoustic evoked responses was highly dependent on the level of vigilance. Its amplitude and duration gradually increased during the progress of slow wave sleep. Parallel registered unit activity showed a marked reduction during the deep-positive slow wave, between 25 and 100 ms. The length of this inhibitory period was in close correlation with the depth of slow wave sleep. Working later with an advanced variation of their cortical multi-microelectrode, Grand (2010) repeated in his Ph.D. work these experiments and found essentially the same: cessation of unit activity during the slow wave specific deep-positive/surface-negative slow component, which he interpreted as "evoked down state" (Fig. 6.4). According to his opinion, "information processing in the cortex during sleep is inhibited by the evoked down states that block the information processing in the 25–100 ms time interval after the stimuli and prevent from awakening if the stimulus is irrelevant. On the global

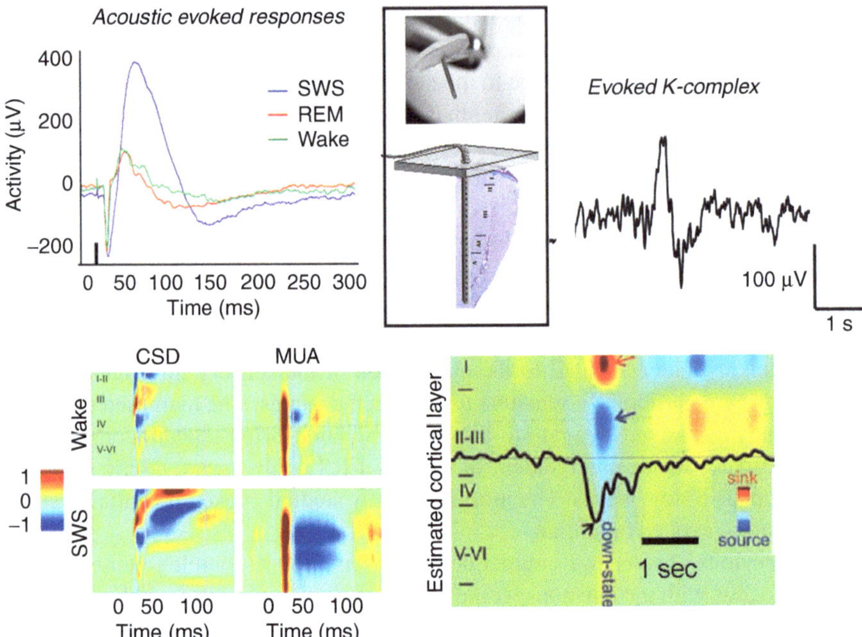

Fig. 6.4 Evoked down states in acoustic evoked responses (cats) and K-complexes (humans). Acoustic evoked responses of cats (Grand 2010) and K-complexes of humans (Cash et al. 2009) were registered by subdural grid and multi-microelectrode (thumbtack) with 24 contacts at 150 µm intervals (*upper part of the insert*). The track of the electrode across the cortex is shown in the *lower part of the insert*. Colored maps show intracortical distribution of current and unit activity according to cortical layers. *Red* indicates inward currents (sink) in CSD and increase in neuronal firing in MUA. *Blue* shows outward currents (source) in CSD and decrease in neuronal firing in MUA. *Black arrow* indicates decreased neuronal firing in the case of evoked K-complex. *SWS* slow wave sleep, *CSD* current source density, *MUA* multiunit activity

cortical scale, human K-complexes act similarly as the evoked down states in the cat auditory system, implicating that sleep protecting mechanisms may essentially be uniform across species."

Congruent with the results of Riedner et al. (2011), this work supports the possibility that sensory evoked potentials, and among them especially those evoked by acoustic stimuli, together with a surface-negative slow wave component, exhibits a stop of cortical cellular activity during NREM sleep. This period of interrupted firing lasts in the acoustic cortex from 25 to 100 ms after the stimulus. This component probably reflects an entrainment of the thalamic inhibitory network and by the thalamocortical and cortico-cortical recruitment probably contributes to the development of the 350–550 negative slow component of K-complex on the level of the macro-EEG.

Congruently, the studies detecting poststimulus time histograms of neuronal firing and cortical ERPs together have found cessation of neuronal activity during the large surface-negative wave component parallel with the down state features of the largest component of K-complex (Cash et al. 2009; Fig. 3.4).

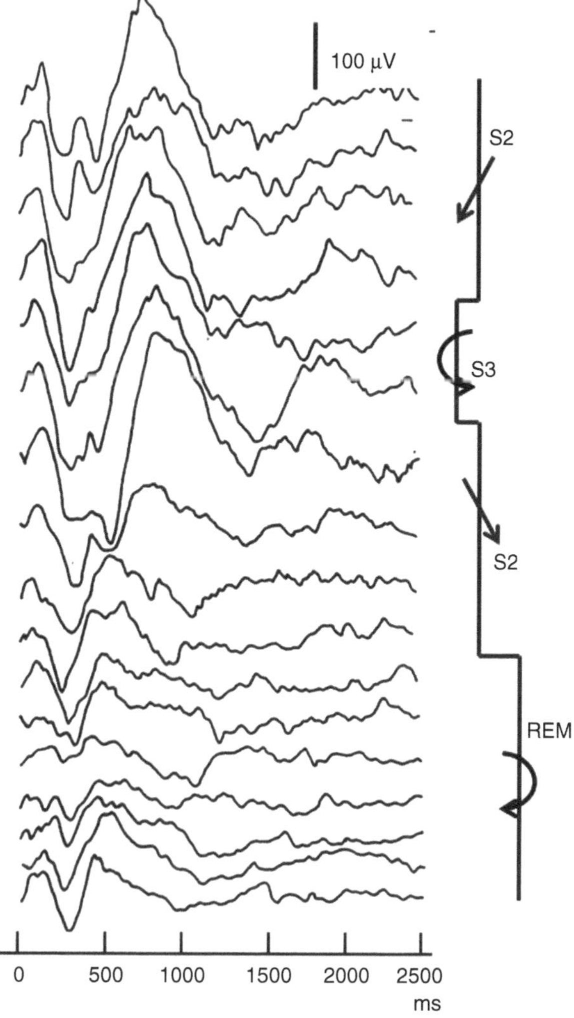

Fig. 6.5 Averaged late components of acoustic evoked responses faithfully follow the cyclic changes of sleep depth. Amplitudes of N350 and P900 are conspicuously higher during the descending slope of the cycle

Another important aspect is the modality independence of the late components and their sensitivity to the sleep EEG environment. In one of our early works (Rajna et al. 1983), the averaged late components of acoustic evoked responses faithfully followed the cyclic changes of sleep depth: their amplitude increased and became more polyphasic as sleep deepened, and decreased as sleep became more superficial. However, the amplitudes proved to be sleep stage independent; rather, it depended only from the position of the given sleep stage in the cycle (Fig. 6.5). Here, we have the opportunity to trace how slow components of ERP (otherwise known as K-complex) in sleep became the building stones of sleep slow oscillation.

This nonmodality specific part of ERP during NREM sleep seems to be part of input-dependent (reactive) slow wave responses known as K-complex or CAP A_1

phase. Here, we have an important convergence of sensory impulses conveying the external world to the brain during sleep and the slow wave elements of NREM sleep fueled by these sensory impulses.

As we have seen, K-complex can be considered a model of the relationship between evoked potentials and sleep slow waves as a reactive (arousal related) form of slow waves. The visibility of input dependency compared to the later continuous slow wave activity in deeper NREM sleep is more obvious. At this stage of sleep when K-complexes are most prominent, the sleep process is not in an advanced stage. When an arousal input comes, the ambiguous elaboration of the "wake or not to wake" question is reflected by the Janus-faced K-complex. In this regard, the findings of Jahnke et al. (2012) are especially relevant. They showed an activation of the so-called default mode network (DMN) (precuneus, anterior cingulate cortex, inferior parietal lobule, and prefrontal cortex) during K-complex and interpreted it in the framework of the "sentinel" hypothesis. According to this concept, DMN represents a special working mode of the brain, in which attention is not focused to anything particular, but the low level floating attention (subconscious state) plays the role of a sentinel, monitoring the external world for unexpected events, also for the protection of sleep.

6.3 Artificial Boosting Slow Oscillation During Sleep (Studying the Basic Slow Wave Producing Defense Mechanism)

Boosting of sleep slow waves has its early history, but it has received much more attention recently (Fig. 6.6). The enhancement of slow wave sleep and of slow waves has its roots in studies assessing the effects of presleep passive body heating on subsequent sleep. It was repeatedly shown that even a warm bath (1 h immersion into 41 °C water) before going to sleep can significantly enhance the time spent in slow wave sleep, especially the time of stage 4 sleep. Quantitative EEG measures of delta activity were also increased. The effects were counteracted by aspirin and shown to decrease rapidly when the time of day of the warm bath receded from the onset of sleep (Bunnell et al. 1988; Raymann et al. 2008; Horne and Shackell 1987). Moreover, even the slow wave sleep enhancement after daytime physical activity was shown to be mediated – among other factors – by the increase in body temperature (Horne and Moore 1985).

Another early root of boosting sleep slow waves was described by Halász and Ujszászi (1991) using acoustic stimulation. "The poststimulus spectral pattern is characterized by a short initial power elevation and a following reduction of all frequency bands except a simultaneous but prolonged (5–20 s) and strong (50 %) power reduction at the 13–14 Hz sigma spindle band. This phenomenon seems to be a common feature in different stages of slow wave sleep" (Halász 1993).

There are several pharmacological agents producing a significant increase in sleep slow wave activity of humans and other mammalian species (Table 6.1). Among these the serotonin receptor 2 (5-HT_{2A}) antagonists are of special interest, because of their use in the treatment of insomnia (Walsh 2009).

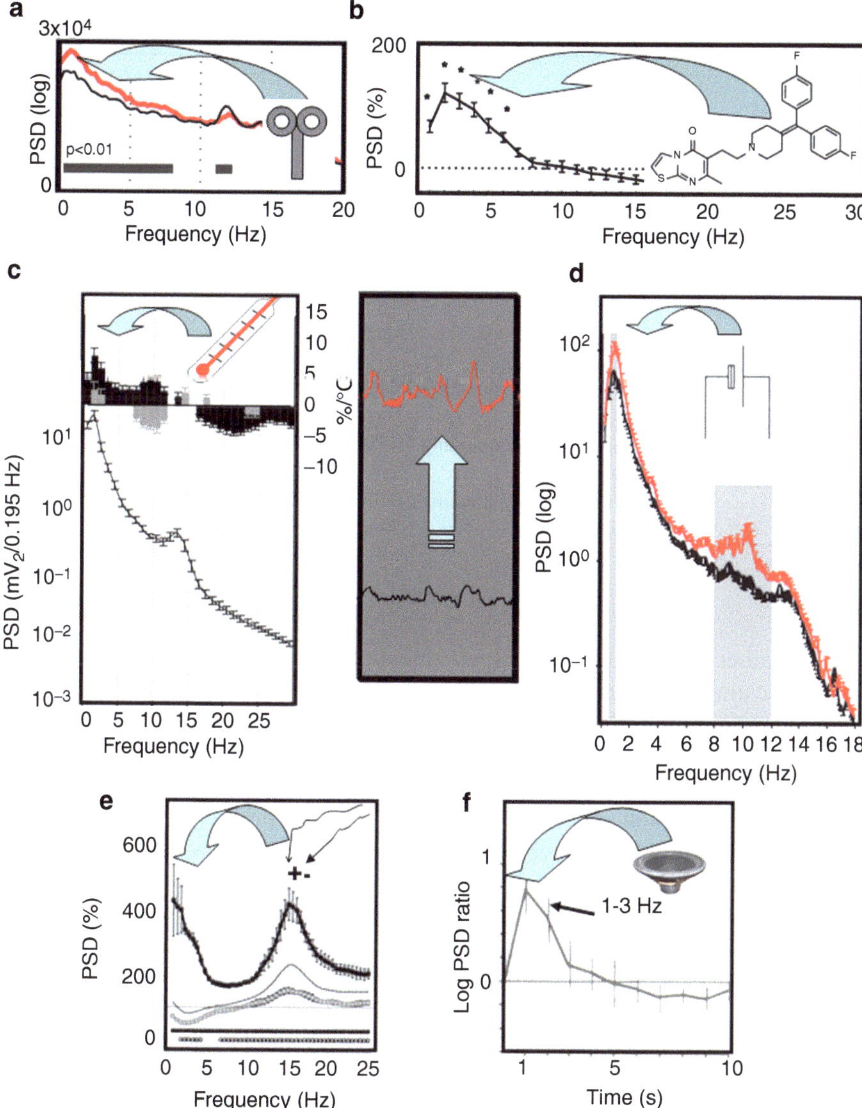

Fig. 6.6 Boosting of slow waves during NREM sleep with different methods increases the spectra of slow wave activity by the intervention. (**a**) Transcranial magnetic stimulation: red line indicates stimulation-induced elevation in spectra of slow wave activity (Massimini et al. 2007). (**b**) Ritanserine a 5-HT2A/2C antagonist: percent increase in slow wave spectra (Kantor et al. 2002). (**c**) Passive body heating (Raymann et al. 2008). (**d**) Direct current stimulation: *red line* indicates stimulation-induced elevation in spectra of slow wave and sigma activity (Marshall et al. 2006). (**e**) Intracortical electrical stimulation (Vyazovskiy et al. 2009). (**f**) Frequent sub threshold acoustic stimulation: log ration of post/pre-stimulation EEG power in the 1–3 Hz range (Halász and Ujszászi 1991)

Table 6.1 Drugs known to increase slow wave sleep

Drug	Mechanism of action	Reference
Tiagabine	GAT-1 inhibitor	Mathias et al. (2001)
Gaboxadol	Selective extrasynaptic GABAA agonist	Deacon et al. (2007)
Gabapentin	A_2-δ site on voltage-gated calcium ion channels	Bazil et al. (2005)
Pregabalin	A_2-δ site on voltage-gated calcium ion channels	Hindmarch et al. (2005)
GHB	GABAB/GHB agonist	Pardi and Black (2006)
Ritanserin	Partially selective 5HT2A receptor antagonist	Dahlitz et al. (1990)
Eplivanserin	Antagonist of serotonin two A receptors (ASTAR)	Hindmarch et al. (2008)
Mirtazapine	Multiple receptors, including 5HT2 antagonist	Shen et al. (2006)
Olanzapine	Multiple receptors, including 5HT2 antagonist	Sharpley et al. (2005)
Trazodone	Multiple receptors, including 5HT2 antagonist	Mendelson (2005)

After Walsh (2009)

Marshall et al. (2006) induced slow oscillation-like potential fields by transcranial application of polarizing current through bilateral frontolateral (anode) and mastoid (cathode) electrodes, oscillating at a frequency of 0.75 Hz during early nocturnal NREM sleep. The artificially induced slow oscillation increased slow wave sleep immediately and entrained slow spindle activity as well over the frontal cortex. The stimulation enhanced retention of hippocampus-dependent declarative memories of healthy human subjects. They concluded that slow oscillation per se plays a role in sleep-associated consolidation of memory, and boosting slow oscillation by externally applied current enhances this process.

Massimini et al. (2007) were able to trigger sleep slow waves resembling spontaneously occurring ones by TMS. These slow waves were harboring spindle activity similarly to the surface-negative phase of the spontaneous slow activity. The well-known traveling properties of these waves were similar to their spontaneous counterpart as well. The slow waves produced by stimulation resulted in spread to the adjacent cortical areas. Compared with K-complexes, the evoked slow waves had much shorter latencies (300 ms or less) and were elicitable by each stimulus with high triggering probability (0.95). This phenomenon proved to be state dependent and have been elicited exclusively during NREM sleep and not in wakefulness. At the same time, the stimulation importantly increased subsequent slow wave sleep. Applying the stimulation during stage 2, the activity quickly switched to stage 4.

Lastly, Vyazovskiy et al. (2009) studied the effect of intracortical electrical stimulation by brief (0.1 ms) pulses during sleep in rats. The stimuli reliably triggered local field potentials that were indistinguishable from naturally occurring slow waves. They were followed by sleep spindles, and in early sleep compared to late sleep, they were larger, had steeper slope, and had lesser peaks. In other words, they were quite similar to the spontaneous slow waves. If the stimulus was delivered immediately after a large spontaneous slow wave, there was no response at all. An important finding of this study was that parameters of evoked slow waves change as a function of the homeostatic sleep pressure. In the condition of increased sleep pressure, the sleep responses were more abundant, and also the slopes of the evoked slow waves were steeper.

The amplitude of evoked slow waves and the incidence of spontaneous high-amplitude slow waves during a NREM period showed reciprocal relationship: when the incidence of high-amplitude slow waves was lowest (at the beginning of the NREM cycle), the amplitude of evoked slow waves were highest, and when the incidence of spontaneous slow waves was highest (at the end of the cycle), the amplitude of evoked slows was the lowest. In other words, the amplitudes of evoked slow waves changed according to the homeostatic process. The higher the homeostatic pressure, the higher the amplitude of evoked slow waves. This is very much congruent with the relationship of CAP A_1 phase rate with the homeostatic process (Terzano et al. 2000), showing exponential decay from the first to later cycles and similar to the decay of spontaneous occurrence rate of K-complexes from cycle to cycle during the sleep process (Halász 2005).

Exploring cortical reactivity during presurgical evaluation of epileptic patients, single-pulse cortical electrical stimulation (CES) was performed via subdural electrodes during NREM sleep (Entz et al. 2007). After initial biphasic activation, a long-lasting negative ECoG potential (down state) and rebound positivity (up state) was obtained in the close (1–2 cm) vicinity of the stimulation electrode, regardless of the distance from the epileptogenic zone. The layer structure and multiunit activity (MUA), current source density (CSD), and time-frequency activity, the distribution of spectral power changes over time recorded by the intracortical multielectrode after cortical electrical stimulation in a 1,000 ms time span, were analyzedThe pattern of the response was very much similar to pattern recorded during sleep slow oscillation. (Fig. 6.7).

Beyond the heuristic value of reactive slow waves, the paradigms for eliciting slow wave activity (local sleep) that may evolve progressively to generalize and deepen slow wave sleep probably can be used in the practice to enhance sleep-related learning or to improve insufficient sleep (therapy of insomnia) in the near future.

All these results show that during NREM sleep, there is a basic slow wave producing defense mechanism (K-complex, or CAP A_1 phase) that can be released both by physiological sensory stimuli as body heating and acoustic stimulation and by artificial brain stimulation – TMS, polarizing current, or direct cortical stimulation. This input-dependent stereotype and modality independent response could be considered as the elementary building stone of NREM sleep.

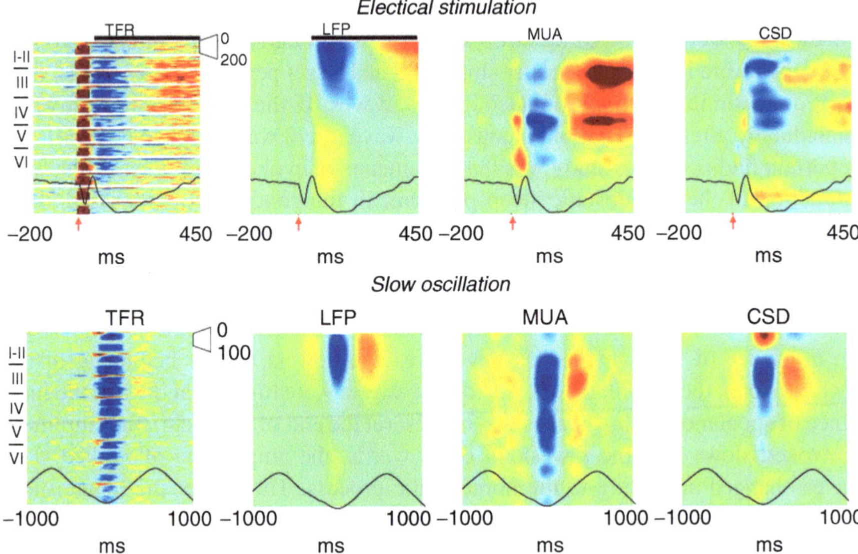

Fig. 6.7 Electrical stimulation-induced local cortical down states and their similarity with spontaneous cortical down states of the slow oscillation during NREM sleep. The figure is the result of co-registered multiunit activity and current source density with an intracortical multi-microelectrode in human subjects. *Top row*: Intracortical responses evoked by electrical stimulation (*red arrow*) of the brain during NREM sleep in an epileptic patient. *Bottom row*: slow oscillation. *TFR*: time-frequency analysis revealing laminar spectral power changes (0–200 Hz) over time after cortical electrical stimulation (*blue*, power decrease; *red*, power increase). *LFP*: Laminar local field potential changes (*blue*, negativity; *red*, positivity). *MUA*: Laminar multiunit activity changes over time (*blue*, decrease in activity; *red*, increase in activity). *CSD*: Laminar current source density (*blue*, source; *red*, sink) (Modified from Entz et al. (2007))

References

Amzica F, Steriade M. Cellular substrates and laminar profile of sleep K-complex. Neuroscience. 1998;82(3):671–86.

Bazil CW, Battista J, Basner RC. Gabapentin improves sleep in the presence of alcohol. J Clin Sleep Med. 2005;1(3):284–7.

Borbély AA. A two process model of sleep regulation. Hum Neurobiol. 1982;1(3):195–204.

Bunnell DE, Agnew JA, Horvath SM, Jopson L, Wills M. Passive body heating and sleep: influence of proximity to sleep. Sleep. 1988;11(2):210–9.

Cash SS, Halgren E, Dehghani N, Rossetti AO, Thesen T, Wang C, Devinsky O, Kuzniecky R, Doyle W, Madsen JR, Bromfield E, Eross L, Halász P, Karmos G, Csercsa R, Wittner L, Ulbert I. The human K-complex represents an isolated cortical down-state. Science. 2009;324 (5930):1084–7.

Coenen AM. Neuronal activities underlying the electroencephalogram and evoked potentials of sleeping and waking: implications for information processing. Neurosci Biobehav Rev. 1995;19(3):447–63.

Dahlitz M, Wells P, James R, Idzikowski C, Parkes JD. Treatment of insomnia with ritanserin. Lancet. 1990;336(8711):379.

Davis H, Davis PA, Loomis AL, Harvey EN, Hobart G. Electrical reactions of the human brain to auditory stimulation during sleep. J Neurophysiol. 1939;2:500–14.

De Gennaro L, Ferrara M, Spadini V, Curcio G, Cristiani R, Bertini M.The cyclic alternating pattern decreases as a consequence of total sleep deprivation and correlates with EEG arousals. Neuropsychobiology. 2002;45(2):95–8.

Deacon S, Staner L, Staner C, Legters A, Loft H, Lundahl J. Effect of short-term treatment with gaboxadol on sleep maintenance and initiation in patients with primary insomnia. Sleep. 2007;30(3):281–7.

Entz L, Fabó D, Erőss L, Halász P, Wittner L, Csercsa R, Dombóvári B, Karmos G, Ulbert I. Detection of slow oscillation detected by cortical electrical stimulation in different vigilance states in humans. In: Society for neuroscience, annual meeting, San Diego, 2007. p. 723.12.

Ferri R, Rundo F, Bruni O, Terzano MG, Stam CJ. Dynamics of the EEG slow-wave synchronization during sleep. Clin Neurophysiol. 2005;116(12):2783–95.

Ferri R, Rundo F, Bruni O, Terzano MG, Stam CJ.Regional scalp EEG slow-wave synchronization during sleep cyclic alternating pattern A1 subtypes.Neurosci Lett. 2006 1;404 (3):352–7.

Gao L, Meng X, Ye C, Zhang H, Liu C, Dan Y, Poo MM, He J, Zhang X. Entrainment of slow oscillations of auditory thalamic neurons by repetitive sound stimuli. J Neurosci. 2009;29 (18):6013–21.

Grand LB. Development, testing and application of laminar multielectrodes and biocompatible coatings for intracortical applications. Ph.D. dissertation, Pázmány Péter Catholic University, Budapest; 2010.

Halász P. Arousals without awakening – dynamic aspect of sleep. Physiol Behav. 1993;54(4): 795–802.

Halász P. K-complex, a reactive EEG graphoelement of NREM sleep: an old chap in a new garment. Sleep Med Rev. 2005;9(5):391–412.

Halász P. The role of the non-specific sensory activation in sleep regulation and in the pathomechanism of generalized epilepsy with generalized spike-wave discharge. Doctoral thesis, The Hungarian Academy of Sciences, Budapest; 1982.

Halász P, Ujszászi J. Spectral features of evoked microarousals. In: Terzano MG, Halász P, Decklerck AC, editors. Phasic events and dynamic organization of sleep. New York: Raven Press; 1991. p. 85–100.

Halász P, Terzano M, Parrino L, Bódizs R. The nature of arousal in sleep. J Sleep Res. 2004;13(1):1–23.

Hindmarch I, Dawson J, Stanley N. A double-blind study in healthy volunteers to assess the effects on sleep of pregabalin compared with alprazolam and placebo. Sleep. 2005;28(2): 187–93.

Hindmarch I, Saubadu S, Delfolie A, Martinez JM, Pinquier JL. Sleep and psychomotor performance in healthy subjects after morning or evening administration of eplivanserin, a novel sleep compound. J Sleep Res. 2008;17(Suppl.1): Abstract No. P359.

Horne JA, Moore VJ. Sleep EEG effects of exercise with and without additional body cooling. Electroencephalogr Clin Neurophysiol. 1985;60(1):33–8.

Horne JA, Shackell BS. Slow wave sleep elevations after body heating: proximity to sleep and effects of aspirin. Sleep. 1987;10(4):383–92.

Jahnke K, von Wegner F, Morzelewski A, Borisov S, Maischein M, Steinmetz H, Laufs H. To wake or not to wake? The two-sided nature of the human K-complex. Neuroimage. 2012;59(2): 1631–8.

Kantor S, Jakus R, Bodizs R, Halasz P, Bagdy G. Acute and long-term effects of the 5-HT2 receptor antagonist ritanserin on EEG power spectra, motor activity, and sleep: changes at the light-dark phase shift. Brain Res. 2002;943(1):105–11.

Karmos G, Molnár M, Csépe V, Winkler I, Grinchenko YV. Evoked cortical fi eld potential and unit responses in the wakefulness-sleep cycle in cats. In: Koella WP, Obál F, Schulz H, Visser P, editors. Sleep '86. Stuttgart/New York: Gustav Fischer Verlag; 1988. p. 287–9.

Marshall L, Helgadóttir H, Mölle M, Born J. Boosting slow oscillations during sleep potentiates memory. Nature. 2006;444(7119):610–3.

Massimini M, Rosanova M. Mariotti M EEG slow (approximately 1 Hz) waves are associated with nonstationarity of thalamo-cortical sensory processing in the sleeping human. J Neurophysiol. 2003;89(3):1205–13.

Massimini M, Ferrarelli F, Esser SK, Riedner BA, Huber R, Murphy M, Peterson MJ, Tononi G. Triggering sleep slow waves by transcranial magnetic stimulation. Proc Natl Acad Sci USA. 2007;104(20):8496–501.

Mathias S, Wetter TC, Steiger A, Lancel M. The GABA uptake inhibitor tiagabine promotes slow wave sleep in normal elderly subjects. Neurobiol Aging. 2001;22(2):247–53.

Mendelson WB. A review of the evidence for the efficacy and safety of trazodone in insomnia. J Clin Psychiatry. 2005;66(4):469–76.

Nicholas CL, Trinder J, Colrain IM. Increased production of evoked and spontaneous K-complcxes following a night of fragmented sleep. Sleep. 2002;25:882–7.

Pardi D, Black J. Gamma-hydroxybutyrate/sodium oxybate: neurobiology, and impact on sleep and wakefulness. CNS Drugs. 2006;20(12):993–1018.

Parrino L, Ferri R, Bruni O, Terzano MG. Cyclic alternating pattern (CAP): the marker of sleep instability. Sleep Med Rev. 2012;16(1):27–45.

Peszka J, Harsh J. Effect of sleep deprivation on NREM sleep ERPs and related activity at sleep onset. Int J Psychophysiol. 2002;46(3):275–86.

Plihal W, Weaver S, Mölle M, Fehm HL, Born J. Sensory processing during early and late nocturnal sleep. Electroencephalogr Clin Neurophysiol. 1996;99(3):247–56.

Rajna P, Halász P, Kundra O, Pál I. Event-related non-specific responses (K-complexes) during sleep. Acta Med Hung. 1983;40(1):33–40.

Raymann RJ, Swaab DF, Van Someren EJ. Skin deep: enhanced sleep depth by cutaneous temperature manipulation. Brain. 2008;131(Pt 2):500–13.

Riedner BA, Hulse BK, Murphy MJ, Ferrarelli F, Tononi G. Temporal dynamics of cortical sources underlying spontaneous and peripherally evoked slow waves. Prog Brain Res. 2011;193: 201–18.

Rosanova M, Timofeev I. Neuronal mechanisms mediating the variability of somatosensory evoked potentials during sleep oscillations in cats. J Physiol. 2005;562(Pt 2):569–82.

Saper CB, Chou TC, Scammell TE. The sleep switch: hypothalamic control of sleep and wakefulness. Trends Neurosci. 2001;24(12):726–31.

Schabus M, Dang-Vu TT, Heib DPH, Boly M, Desseilles M, Vandewalle G, et al. The fate of incoming stimuli during NREM sleep is determined by spindles and the phase of the slow oscillation. Front Neurol. 2012;3:1–11.

Sharpley AL, Attenburrow ME, Hafizi S, Cowen PJ. Olanzapine increases slow wave sleep and sleep continuity in SSRI-resistant depressed patients. J Clin Psychiatry. 2005;66(4):450–4.

Shen J, Chung SA, Kayumov L, Moller H, Hossain N, Wang X, Deb P, Sun F, Huang X, Novak M, Appleton D, Shapiro CM. Polysomnographic and symptomatological analyses of major depressive disorder patients treated with mirtazapine. Can J Psychiatry. 2006;51(1):27–34.

Terzano MG, Mancia D, Salati MR, Costani G, Decembrino A, Parrino L. The cyclic alternating pattern as a physiologic component of normal NREM sleep. Sleep. 1985;8(2):137–45.

Terzano MG, Parrino L, Fioriti G, Spaggiari MC, Anelli S, Arcelloni T, Orofiamma B, Farolfi A. The cyclic alternating pattern rate: a sleep variable sensitive to environmental modifications. A controlled study utilizing white noise and zolpidem. In: Sauvanet JP, Langer SZ, Morselli PL, editors. Imidazopyridines in sleep disorders. New York: Raven Press; 1988. p. 219–30.

Terzano MG, Parrino L, Boselli M, Smerieri A, Spaggiari MC. CAP components and EEG synchronization in the first 3 sleep cycles. Clin Neurophysiol. 2000;111(2):283–90.

Terzano MG, Parrino L, Smerieri A, Carli F, Nobili L, Donadio S, Ferrillo F. CAP and arousals are involved in the homeostatic and ultradian sleep processes. J Sleep Res. 2005;14(4):359–68.

Vyazovskiy VV, Faraguna U, Cirelli C, Tononi G. Triggering slow waves during NREM sleep in the rat by intracortical electrical stimulation: effects of sleep/wake history and background activity. J Neurophysiol. 2009;101(4):1921–31.

Walsh JK. Enhancement of slow wave sleep: implications for insomnia. J Clin Sleep Med. 2009;5 (2 Suppl):S27–32.

Chapter 7
Slow Wave Activity as Substrate of Homeostatic Regulation

Abstract During the last 15–20 years, a new knowledge accumulated about NREM slow wave oscillations that have become the key issue of homeostatic regulation. A frequency-based classification of slow waves has been developed, differentiating between 0.1–1- and 1–4-Hz groups. The cortical <1-Hz slow activity during the so-called up states (surface-positive half wave), even ripple-like (50–200 Hz) fast activity, and down state (surface-negative half wave), an interruption of synaptic and neural activity, have been described. The alternation of these two microstates ensures a unique double working mode for the cortex, providing continuity for the contact and information processing with the environment during the up states even in deep sleep and providing a separation for trophotropic functions for further cognitive demands during the down states.

With progressive development of neuroimaging, source modeling studies on sleep slow waves by new neuroimaging tools have confirmed that the cortical areas are differentially involved in slow wave production and showed that sleep slow waves can be locally – mainly frontally – regulated. They are traveling along an anterior-posterior axis largely mediated by the so-called cingulate highway. Studies in this field emphasized that those areas with maximal involvement in slow waves' production also show considerable overlap with the default network, paradoxically implicated in monitoring the external environment, and can be altered by sleep deprivation.

Ontogenetic studies revealed that the delta oscillation associated with rapid spindling is the agent of plastic changes of the cortex. Reactive (input-dependent) delta activity seems to be an essential element of plastic changes as early as during the neonatal development. Before the fetal brain might receive elaborated sensory inputs from the external word, spontaneous fetal movements provide sensory stimulation and drive delta-brush oscillation, contributing to the formation of cortical body maps.

The spectral power of sleep slow wave activity and the steepness of the slopes of sleep slow waves were shown to correlate positively with the gray matter volume of several cortical areas in children and adolescents between 8 and 19 years of age. When the production of cortical synapses is more efficient than their elimination

P. Halász, R. Bódizs, *Dynamic Structure of NREM Sleep*,
DOI 10.1007/978-1-4471-4333-8_7,

(from birth until the prepubertal age), slow wave activity is high and increasing; while in adulthood, when the elimination of synapses exceeds their production, the amount of sleep slow wave activity decreases.

Discussing phylogenetic relations of slow wave activity during different vigilance states and state-dependent reactions to sensory inputs, we try to interpret some paradoxical observations on reptiles. We are proposing that the reason why reptiles are in a continuous NREM sleep like condition during behavioral waking state is the lack or underdevelopment of their cholinergic arousal system. Therefore, sensory stimulation elicits K-complex-like slow wave responses. In the waking state, reptiles apparently have sleep EEG and sleep-like EEG activity during behavioral activation. Our proposal incorporates an explanation for the lack of long-term homeostatic sleep regulation in reptiles, having at the same time short-term homeostatic slow wave supplementation in response to sensory stimulation.

Keywords Slow oscillation • Delta activity • Infraslow oscillation • Up states Down states • Neuroimaging • Ontogenesis of sleep • Brain development Phylogenesis of sleep • Reptilian wakefulness

7.1 A Short History of Slow Waves

The main focus of our present work is the phenomenon of slow waves, considered as a quintessence of sleep. Slow waves are also known as delta waves according to the classical EEG terminology, originating from William Gray Walter (1910–1977) in the 1930s (Walter 1936). He has performed their topographic mapping and recognized the link between brain lesions and local delta activity (meaning disease, degeneration, and death). The strong link between slow waves and sleep was first reported, also in the 1930s, by Alfred Lee Loomis (1935). He managed to make all-night EEG records of human sleep in his laboratory. This was an unbelievable technical achievement at those times as even Harvard researchers were unable to perform such records. Loomis characterized the EEG stages of sleep: the stage characterized by slow waves was called random, as the waves were considered as such in his view. He also described the increase in the amplitude of delta waves during the course of wake-sleep transition until reaching deep sleep, and most importantly he also recognized the reactivity of the slow waves and made a correct description on K-complexes evoked by knocking on the sleepers' door (K-knock).

Thus, slow waves were considered mainly as signs of low arousal level (Loomis et al. 1935) or brain lesion (Walter 1936). This idea was present in the work of Giuseppe Moruzzi (1910–1986) and Horace Winchell Magoun (1907–1991) also who described the ascending reticular activating system in the brainstem and characterized it as follows: "The evidence given above points to the presence in the brain stem of a system of ascending reticular relays whose direct stimulation activates or desynchronizes the EEG, replacing high-voltage slow waves with low-voltage fast activity. This effect is exerted generally upon the cortex and is mediated, in part, at least, by the diffuse thalamic projection system" (Moruzzi and Magoun 1949). High-voltage slow waves were already considered as signs of sleep or low arousal in these works. This kind of "negative" view on slow waves (i.e., being signs of lack of something) persisted and was even more emphasized during and after the discovery of REM sleep in humans (Aserinsky and Kleitman 1953) or paradoxical sleep in cats (Jouvet et al. 1959). It was the appearance of fast waves instead of slow ones during sleep that seemed interesting in those times.

However, scientific interest has changed during the next few years. The search for the molecular sleep factor (S-factor) by the Pappenheimer group stimulated the search for some reliable physiological measure of sleep deprivation-induced sleep rebound. It was recognized that the best measure is sleep EEG slow wave activity. They wanted to introduce into these studies the time spent in those sleep phases characterized by EEG slow waves as well as the increase in their amplitude, instead of focusing on low-voltage fast activity (Pappenheimer et al. 1975). Similar approaches characterized the research on humoral sleep regulation (Schoenenberger et al. 1977).

Although the homeostatic regulation of slow wave sleep was already known in those studies, it was Wilse B. Webb who explicitly stated that the amount of stage 4 sleep was related to the amount of pre-sleep wakefulness (Webb and Agnew 1971). Hints on the potential recuperative role of slow wave sleep together with its precise homeostatic regulation changed the view of sleep researchers and have led to the generally accepted theories: slow waves are not merely passive expressions of reduced cortical activity due to brain lesions or low arousal; rather, they represent active self-organizational forms of neuronal activity, serving recuperative and off-line information processing functions.

7.2 Is There a Frequency-Based Typology of Slow Waves?

7.2.1 *The Controversy of Below or Above 1 Hz*

Mircea Steriade and his co-workers published a series of papers (Steriade et al. 1993a, b, c) reporting that:

1. Cellular processes contributing to delta waves (<4 Hz) are inhomogeneous.
2. The 0.1–1-Hz component of the delta (<4 Hz) activity has distinct cellular substrates and is of utmost importance in sleep rhythm generation.
3. The 0.1–1-Hz waves are reflections of large-scale, rhythmic hyperpolarizations followed by widespread depolarization.
4. The hyperpolarization-depolarization sequences originate from cortical neurons and are synchronized by corticocortical connections.

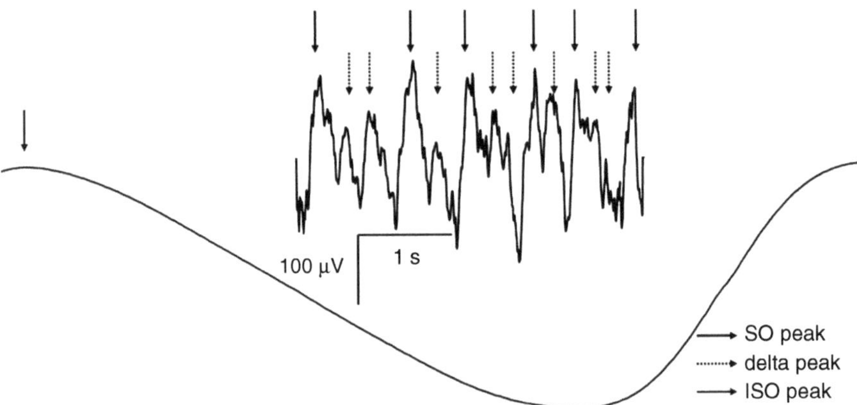

Fig. 7.1 Infraslow (0.01–0.1 Hz) oscillation (*ISO*), slow (<1 Hz) oscillation (*SO*), and delta (1–4 Hz) activity are coexisting and coalescent during slow wave sleep in humans. Negative peaks of the ISO, SO, and delta waves are shown by *arrows*. SO waves and delta waves emerge during the positive going phases (up states) of the ISO and SO, respectively

5. The 0.1–1-Hz waves envelope delta and sigma (spindle) oscillations, resulting in complex, coalescent wave sequences (Fig. 7.1).

This novel cortical rhythm was called the slow oscillation, and the "<1-Hz" symbol was usually added in parentheses in order to make the distinction clear between the slow and the delta rhythms. According to this view, the 1–4-Hz waves are reflections of the thalamic clock-like delta activity and the cortical delta activity, while the <1-Hz component is of cortical origin reflecting different physiological processes (Amzica and Steriade 1998). A detailed scenario of sleep rhythm generation in the thalamocortical system was also given. This scenario was based on the hypothesis that the hyperpolarization-rebound sequences of thalamocortical feedback loops generating the spindle and the delta waves are triggered and grouped by the depolarization phases (up states) of the slow oscillation (Steriade et al. 1993c). The distinction between the slow oscillation and the delta waves was confirmed by the sleep records performed in animal model with the lack of T-type Ca^{2+} channels. The lack of this channel resulted in a significant decrease in delta and spindle oscillations, but not the slow (<1 Hz) oscillation, which remained unaffected (Lee et al. 2004). Other molecular evidences for the differentiation between the slow oscillation and the delta waves have come from pharmaco-EEG studies of sleep. Benzodiazepine hypnotics decrease NREM sleep EEG power in the delta range (>1 Hz), but may significantly increase the slower (<1 Hz) frequency components (Trachsel et al. 1990). Moreover, only the EEG frequencies <1.5 Hz, but not the higher delta bins, are affected by noradrenaline depletion in sleep-deprived rats: neurotoxic lesions with DSP-4 reduce the 0.5–1.5-Hz activity in recovery sleep, while >1.5-Hz activity remains unchanged (Cirelli et al. 2005).

Some years after the first description of the slow oscillation in animals, it was shown that its role in human sleep homeostasis is different from that of delta activity. The fast Fourier transformation (FFT) power of the <1-Hz component does not significantly decrease between the first and the second sleep cycles, while the power of the 1–4-Hz delta component does (Achermann and Borbély 1997). This was interpreted as a sign for the lack of the homeostatic regulation of the slow oscillation. The idea on missing homeostatic regulation of the slow oscillation was supported by early studies revealing a decrease in NREM sleep EEG power >1 Hz after prolonged periods of wakefulness or sleep deprivation in humans and rats (Borbély et al. 1981; Tobler and Borbély 1990). However, this conclusion was a premature one. Although the time course of delta activity is characterized by a steeper decline over the sleep cycles (NREM periods) than the time course of the slow oscillation, in physiological conditions, both seem to be under precise homeostatic regulation. This was proven by examining the effects of intervening naps in studies performed in young good sleepers (Campbell et al. 2009). Based on their different time courses, Campbell et al. (2009) attributed a permissive role to the slow oscillation in sleep homeostasis: that is, the slow waves permit the expression of the actual sleep need, measurable by the actual level of delta, embedded in sleep cycles (Fig. 7.1). The slow oscillation provides the neural and metabolic conditions for homeostatic processes. The explanation of the peculiar decrease of the <1-Hz NREM sleep EEG in recovery sleep after sleep deprivation (Borbély et al. 1981; Tobler and Borbély 1990) was given by the studies examining the frequency and the slopes of the slow waves under different levels of sleep pressure. These studies have revealed the acceleration in the alternation of up and down states, thus an increase in the frequency of the slow waves as well as an increase in the steepness of slow waves in conditions of increased sleep pressure (Riedner et al. 2007; Bersagliere and Achermann 2010; Fig. 7.2). Several results suggest that the original frequency border of <1 Hz for the slow oscillation is too narrow, and in certain conditions, frequencies higher than 1 Hz may contribute to the phenomenon. Moreover, reports of the decreased <1-Hz activity during recovery sleep find their reason in a frequency shift or acceleration of the slow oscillation. The apparent dissociation between the frequencies >1 Hz and the remaining slow wave activity may simply reflect homeostatic changes in the slopes of slow waves (Hanlon et al. 2011).

The cortical origin and corticocortical synchronization of the slow oscillation have led researchers to find the cognitive correlates of this sleep EEG rhythm. Several evidences were found for the correlation between the spectral measures of the slow oscillation and performance in different cognitive tasks. The performances were found to correlate with the sleep slow oscillation and were related to visuospatial memory as measured by the Rey-Osterrieth Complex Figure Test (Bódizs et al. 2002) and different forms of executive functions like nonverbal planning or word generation (Anderson and Horne 2003), as well as the overnight improvement in motor sequence learning (Moroni et al. 2008). Verbal learning was shown to increase EEG coherence in the up states of the slow oscillation during post-sleep learning (Mölle et al. 2004). Last, but not the least, the artificial boosting of the slow oscillation by applying transcranial direct current stimulation of 0.75-Hz frequency has significantly increased the overnight retention of paired associate verbal material (Marshall et al. 2006). It seems that there are two main cognitive features of the slow oscillation which were unraveled by the above investigations. One is a link

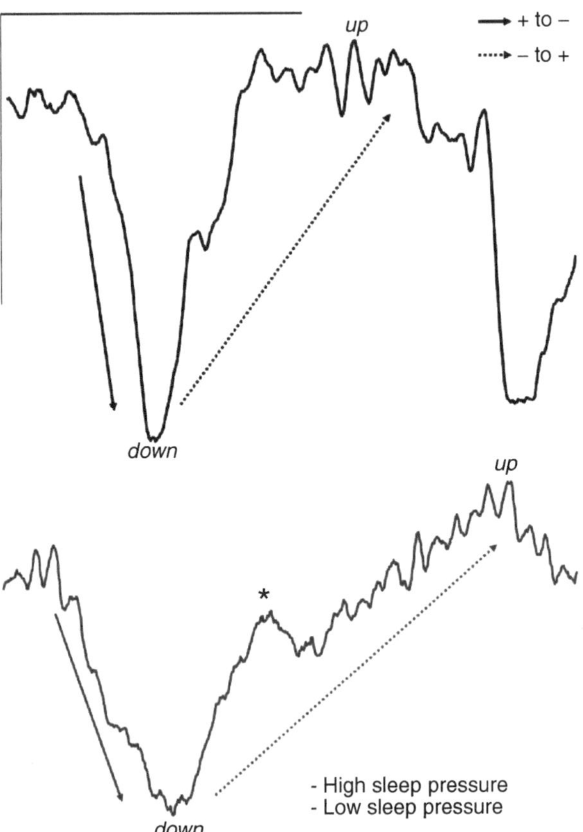

Fig. 7.2 General morphology, up and down states, as well as sleep pressure dependence of the slow (<1 Hz) oscillation in human slow wave sleep EEG. Down states are characterized by surface negativity and relative lack of superimposed higher-frequency oscillations, whereas up states are surface positive and superimposed with spindling (sigma activity). Higher sleep pressure is characterized by faster alternation of the up and the down states, steeper slopes, higher amplitude, as well as a relative lack of the well-formed multipeaks. The multipeak phenomenon during the lower sleep pressure is shown by an *asterisk* (*). Calibration marks: 100 μV and 1 s. EEG recordings come from stage 3 sleep of the first (high sleep pressure) and the third (low sleep pressure) sleep cycles of a young healthy subject (derivation F4-A_1)

between some type of cognitive performances and the baseline individual level of the slow oscillation. This could reflect a trait-like relationship revealing the commonalities in or reciprocal relationships between the neural background of wakeful cognitive performances and of NREM sleep cortical slow oscillation (Bódizs et al. 2002; Anderson and Horne 2003). The other type of cognitive aspect of the sleep slow oscillation is related to the sleep-dependent mechanisms of memory consolidation. There is strong evidence for the involvement of the sleep slow oscillation in the consolidation of memory traces (Marshall et al. 2006; Moroni et al. 2008).

7.2.2 *Below 0.1 Hz: Infraslow Oscillation in Light of the Full-Band EEG Recordings*

Direct current (DC)-EEG recordings of human sleep revealed a prominent infraslow activity roughly associated with 0.01–0.1-Hz frequency fluctuations (Vanhatalo et al. 2004; Monto et al. 2008; Picchioni et al. 2011; Hughes et al. 2011). Although the exact frequency limits of the infraslow oscillation are arbitrary and vary from study-to-study, the main issue unraveled was the evidence showing that it modulates neural excitability and interictal epileptic activity during sleep (Vanhatalo et al. 2004). Frequencies higher than 0.5 Hz were shown to be phase-coupled with (grouped by) the infraslow oscillation during both wakefulness (Monto et al. 2008) and NREM sleep (Vanhatalo et al. 2004; Fig. 7.1). The prominent very slow oscillatory patterns of preterm infants disclosed by DC-EEG during sleep were shown to take the form of long-lasting (1–5 s) occipital negative transients (200–700 μV), embedding the typical delta bursts seen in the conventional EEG (Vanhatalo et al. 2002). The power of the infraslow oscillation during sleep correlates positively with the BOLD signal intensity measured by fMRI in many subcortical regions (cerebellum, thalamus, basal ganglia) as well as some lateral cortical areas and the hippocampi. In contrast, the BOLD signal intensity of the paramedian heteromodal cortices correlated negatively with the power of infraslow activity (Picchioni et al. 2011). This was interpreted as the organization of the broad dissociation of activity between cortical and subcortical regions proven by the infraslow EEG oscillation during sleep. In vivo and in vitro data suggest that the thalamus could be a source of the infraslow oscillation (Lörincz et al. 2009; Hughes et al. 2011).

A particularly wide range of neurobehavioral phenomena fluctuate at frequencies corresponding roughly to the 0.01–0.1-Hz range. These fluctuations of autonomic nervous system activity (Lambertz and Langhorst 1998) had remained largely unmentioned in previous publications. Accumulating evidence suggests that processes providing the infraslow oscillation are the cellular substrates of some large-scale fluctuations which can appear as the CAP phenomenon in some parts of sleep.

We believe that the periodicity of the micro-arousals embedded in the CAP sequences is in fact due to the infraslow oscillation, shaping gross measures of neuronal excitability and being present in many neurobehavioral phenomena analyzed at this time scale (Table 7.1). The parallelism between these infraslow oscillations is striking when one considers that the periodic fluctuations in arousal (CAP) probably reflect the periodic fluctuations in neural excitability (infraslow oscillation). Thus, the periodicity in arousal could be gated by the periodicity in neural excitability. CAP could be a kind of an enhancement or phase resetting of infraslow activity due to internal or external sources of arousal or other forms of sleep instability. The possible link between the infraslow activity and the CAP phenomenon was suggested by other authors earlier (Lörincz et al. 2009).

7.3 Up and Down States of Slow Waves

We are witnessing that the hitherto very poorly understood sleep EEG graphoelements partially described earlier come step-by-step to life and fill up by meaning. They behave as an organic system with biological role, like the pieces of wood transformed to Pinocchio, a living child with full conscience in the hands of master Geppetto, in the famous tail.

We learned from the Steriade group that different rhythms (spindles, delta, and <1-Hz slow waves on one hand and fast frequencies on the other) prevail in coalescence during slow wave sleep.

Table 7.1 Neurobehavioral phenomena with periodicities in the CAP range

Neurobehavioral phenomena	Species	State	Frequency (Hz)	Period (s)	Specific remarks/conditions	Reference(s)
Burst suppression	Human	Coma or general anesthesia	Highly variable		Cyclic simultaneous variation of EEG patterns	Fishgold and Mathis (1959)
Slow wave (<4.5 Hz) EEG amplitude	Human	NREM sleep	0.031–0.047	21.27–32		Achermann and Borbély (1997)
Theta- and alpha-band power	Human	Wakeful resting	0.024–0.057	17.54–41.6		Novak and Lepicovska (1992)
Ultra-slow rhythm	Human	NREM sleep	0.05–0.025	20–40		McKeown et al. (1998)
Alternating amplitude segments	Rat	NREM sleep	<1	>1		Deporteere et al. (1991)
Periodic oscillations in EEG and behavior	Squirrel monkey	Wakefulness	0.06–0.03	15–30	Chair-restrained, alternation of vigilance and inattentiveness	Ehlers and Foote (1984)
Oscillation in baseline activity of basal ganglia neurons	Rat	Wakefulness	0.5–0.016	2–60		Ruskin et al. (1999)
Very slow oscillatory behavior of lateral geniculate neurons	Rat	Urethane anesthesia or wakefulness	0.01–0.025	40–100	Oscillation blocked by illumination of the eyes, re-induced by (non-) NMDA-antagonists or GABA agonists	Albrecht et al. (1998)

Alternating network excitability of the hippocampus (CA_1 region)	Rat	Urethane anesthesia	0.025	40	Triggers epileptic-like afterdischarges during periods of enhanced excitability	Pentonnen et al. (1999)
Cyclic activations in the brainstem, respiration, cardiovascular system, and EEG	Dog	Chloralose-urethane anesthesia	0.05–0.5	2–20	The rhythms mutually influence their frequencies	Lambertz and Langhorst (1998)
Infraslow EEG oscillation	Human	NREM sleep	0.02–0.2	5–50	DC-EEG recorded fluctuations, grouping higher-frequency EEG rhythms	Vanhatalo et al. (2004)
fMRI BOLD signal fluctuation	Human	Both sleep and wakefulness	<0.1	>10	Fluctuations identify the default mode network (DMN)	Riedner et al. (2007)

Within the cortical <1-Hz slow activity during the so-called up states (surface-positive half wave) even ripple-like (50–200 Hz) fast activity and during the "down states" (surface-negative half wave), an interruption of synaptic and neural activity alternates (Fig. 7.2). "The entire cortical network can swing rhythmically between extremely different microstates, ranging from wakefulness-like network activation to functional disconnection in the space of a few milliseconds" (Massimini et al. 2003). The "black-hole"-like down state may provide a rest phase and restoration for the most vulnerable and recuperation demanding frontal cognitive activity.

In the work of Csercsa et al. (2010) intracortical laminar local field potential gradient, multiple-unit and single-unit activities were recorded during slow wave sleep, related to simultaneous electrocorticography, and were analyzed by current source density and spectral methods. We found that slow wave activity in humans reflects a rhythmic oscillation between widespread cortical activation and silence. Cortical activation was demonstrated as increased wideband (0.3–200 Hz) spectral power including virtually all bands of cortical oscillations, increased multiple- and single-unit activities, as well as powerful inward transmembrane currents, mainly localized to the supragranular layers. Neuronal firing in the "up state" was sparse, and the average discharge rate of single cells was less than expected from animal studies. Synchronization patterns of action potentials across all cortical layers suggested that any layer could initiate firing at the onset of up states.

These findings provide strong direct experimental evidence that slow wave activity in humans is characterized by hyperpolarizing currents associated with suppressed cell firing, alternating with high levels of oscillatory synaptic/transmembrane activity associated with increased cell firing. Our results emphasize the major involvement of supragranular layers in the genesis of slow wave activity.

The recurrent cortical disfacilitation during down states together with high-level cortical activity during up states has a serious consequence for sensory transmission. The main one is that the thalamocortical sensory processing during deep slow wave sleep is not steady (Massimini et al. 2003). This could be shown by experiments where the averaging of sensory evoked potentials was grouped according to the up and down states of <1-Hz slow wave oscillation in deep sleep of animals. By this method, the amplitudes of cortical somatosensory evoked responses were found to be different: high, although not reaching wakefulness-like levels, during the up states (confirming the possibility of detecting meaningful events even in deep sleep) and low during the disfacilitation of the down state. This lack of stationarity may impair sensory integration during deep slow wave sleep.

7.4 Imaging of Slow Waves in Sleep

The activation of certain brain structures in slow wave sleep has been recorded by high-tech neuroimaging methods recently.

The advent of functional neuroimaging allowed the registration of regional cerebral blood flow (SPECT) and metabolic (PET scanning) and later neuronal activity BOLD (fMRI) changes also for sleep studies.

PET studies using/18F/-fluorodeoxyglucose as a tracer showed that global cerebral glucose metabolism in humans is lowest in stages 2–3 in NREM sleep, in contrast to REM sleep in which it is similar to the waking state (Maquet et al. 1990). Later, the resolution of PET methods substantially improved both in time and in space, and a more reliable determination of the regional differences during sleep has become possible (Braun et al. 1997; Hofle et al. 1997; Nofzinger et al. 1997; Maquet and Phillips 1998; Maquet 2000). A significant decrease (greater than in the rest of the brain) in rCBF has been shown during NREM sleep in the brainstem (dorsal pons and mesencephalon), thalamus, basal ganglia, basal forebrain, and hypothalamus and in the orbitofrontal and anterior cingulate cortex, as well as the precuneus.

The fMRI studies offer sophisticated knowledge about BOLD activation patterns related to sleep (Czisch et al. 2002, 2004, 2009; Kaufmann et al. 2006; Schabus et al. 2007; Dang-Vu et al. 2008). Several brain regions' activities were found to be reduced during each NREM sleep stage. Most of these regions are located in the frontal lobes, but some regions of the thalamus, limbic system (anterior and posterior cingulum), as well as wider areas of the temporal, parietal, and occipital lobes, and the midbrain, hypothalamus, and mamillary bodies have also been shown to have decreased activity (Kaufmann et al. 2006). Certain patterns of activation sequences were described along the process of falling asleep, like loss of alertness during expectation of environmental stimuli as well as during deepening sleep. The decrease of insular activity gained significance related to interoceptive awareness. Special attention was paid to the crucial hypothalamic region showing globally decreased activity reflecting to the decreased firing of wake-promoting neurons. It is interesting that "some pronounced synchronous BOLD activity changes" were observed in the hypothalamus in NREM that correlated with a simultaneous activity in wider cortical areas resembling to the activity in the pathways of the ascending arousal system during NREM sleep. Kaufmann et al. (2006) emphasize the role of a hypothalamic-driven network regulation both in the initiation and maintenance of sleep. They observed a decreased activity of the hypothalamic regions throughout NREM sleep stages, reflecting the reduced activity of wake-promoting neurons. Due to limited spatial resolution, differentiating hypothalamic subregions was not possible. However, "some pronounced synchronous BOLD activity changes that correlated with the hypothalamus occurred: limbic structures, regions of the frontal and parietal cortex, the basal forebrain and brainstem showed a similar time course of activation as hypothalamic regions did." Kaufman et al. interpreted this pattern of activation as resembling to the pathway of the ascending arousal system which sends projections from the brainstem and posterior hypothalamus throughout the forebrain. These findings are congruent with the role of arousal-driven phasic activity, as described in this book. Phasic increases were observed in regional brain activity in relation to discrete events as slow waves or delta waves, compared to baseline sleep activity (Dang-Vu et al. 2008). Delta waves (>1 Hz) were associated with frontal responses. Significant increases in activity were associated with the <1-Hz waves in several cortical areas, including the inferior frontal, medial prefrontal, and precuneus and posterior cingulate areas. The <1-Hz slow waves are associated with significant activity in the parahippocampal gyrus, cerebellum, and brainstem, whereas delta waves are related to frontal regions compared to baseline activity. No decrease of activity (negative BOLD signal) was observed. A partial overlap was seen between the default network of wakefulness and responses associated with slow wave oscillation. They unexpectedly found significant responses associated with slow oscillations in the midbrain and pontine tegmentum, especially in regions of cholinergic and aminergic nuclei. This particular finding again points to traces of arousal activity during NREM sleep, the importance of which was highlighted in Chap. 6

Using high-density EEG and source modeling methods, sleep slow waves occurring spontaneously and evoked by transcranial magnetic stimulation were studied (Murphy et al. 2009). It was found that some areas of the cortex are differentially involved in slow waves and that sleep slow waves can be locally regulated. Slow waves as a group were associated with large currents in the medial, middle, and inferior frontal gyri, the anterior cingular, the precuneus, and posterior cingular regions. This study again emphasized that areas showing maximal involvement in slow waves also show considerable overlap with the default network which is paradoxically implicated in monitoring the external environment and can be altered by sleep deprivation.

Further understanding of the role of slow wave oscillation and particularly of the up and down states in plastic changes during sleep would be crucial; however, spatial resolution of fMRI does not allow us to reveal the time and space structure of cortical up and down states during NREM sleep changes.

7.5 Relation of K-Complexes and Sleep Slow Waves

We have shown that K-complexes express a double nature (Janus-faced): they show responsivity to sensory (mainly acoustic) stimuli and behave as slow waves following the homeostatic decay increasing after sleep deprivation and being related to the deepness of the actual sleep cycle. Furthermore, the studies of Amzica and Steriade (1998) and later Cash et al. (2009) proved that the main surface-negative large N550 component of K-complexes is characterized by an important decrease of unit discharges and synaptic activity in the frontal cortex similar to the “down state” of the frontal slow oscillation during NREM sleep (Fig. 3.4).

According to the fMRI studies of Riedner et al. (2011), even the spontaneous K-complexes wear, at their initial segment, the traces of multisensory processing. Therefore, we may consider K-complexes as the most characteristic prototypes of the “reactive slow wave activity.” Although K-complexes can be well obtained by the averaging also in deep slow wave sleep (Fig. 3.3), they are more frequent in stage 2.

K-complexes are essential constituents of CAP sequences, mainly in the A_1 phase events. At the same time A_1 events occur much more frequently during the descending slopes of the first cycles, but both spontaneous and elicited K-complexes have higher frequency during the “A” slope. This difference in distribution is even more pronounced if we consider the behavior of evoked K-complexes under continuous random sensory stimulation (Halász 1982). While less reactive slow activity can be elicited in the form of CAP A_1 phases during the "A" slopes of the first cycles, the responsivity in the form of singular K-complexes is more preserved compared to the "D" slopes.

Thus, K-complex occurs more frequently during stage 2 sleep and during "A" slopes of cycles and seems to be more linked to the arousal compared to other sleep slow waves. If we involve into the group of reactive sleep graphoelements the classical vertex sharp transients, a continuum of these graphoelements can be outlined more clearly. In superficial sleep (stage 1–2), sensory (mainly acoustic) stimuli evoke the vertex sharp wave, which can be considered as an exaggerated form of acoustic evoked potential, without any important slow wave constituent. Later, in stage 2, the stimulation evokes K-complexes instead of vertex waves, with recognizable remnants of the acoustic evoked potential, but the essential part is the slow wave complex (N350-N550-P900, see Fig. 3.3). Beginning from stage 2, in deeper NREM sleep, the evoked response becomes more complex in the form of CAP A_1, which incorporates K-complexes, slow waves, and spindles, while the original evoked potential compartment is hardly visible. The response to sensory stimulation shifts from a multimodal evoked potential-type response to a nonspecific bilateral, frontally dominant slow wave response.

The recent functional neuroimaging results seem to confirm this continuum. Stern et al. (2011), investigating vertex sharp wave generation by fMRI, have shown that BOLD positive activations of regions principally encompass the primary sensorimotor cortical regions for vision, hearing, and touch. Earlier in this book, we detailed the findings of Riedner et al. (2011) who had highlighted the initial evoked potential compartments before the slow wave components of K-complexes. The fMRI counterparts of sleep slow waves were discussed in detail in Sect. 6.2.

The special situation of K-complex in sensory information workup and in fulfilling trophotropic and plastic functions of NREM sleep is nicely supported by the findings of Jahnke et al. (2012). In their fMRI study, they found that K-complex is associated with positive BOLD signal changes in subcortical (brainstem and thalamus), cerebellar, sensory (auditory, visual, and sensorimotor) motor, and midline (anterior and midcingular gyri, precuneus) areas and regions which form part of the so-called default mode network (DMN) (prefrontal cortex, inferior parietal lobule, precuneus). Negative BOLD signal changes were observed in bilateral insular cortices. The primary auditory cortex was the first cortical region to be influenced during K-complex.

These findings confirm "the central role of the thalamus in the mediation of cortically generated K-complex leading to sleep spindles and slow waves" and also that "K-complex permits a low level information processing in one hand and a sleep protective counteraction on the other." The DMN-type activation is interpreted in line with the "sentinel hypothesis," describing DMN activity "in the context of a broad low level focus of attention when one – like a sentinel – monitoring the external world for unexpected events." This version of the Hamletian question ("to wake or not to wake") is very near to our interpretation (Halász 2005).

Table 7.2 Continuum among the reactive graphoelements in NREM sleep

Reactive sleep graphoelements	Vertex wave	K-complex	Slow waves
Reactivity	Spontaneous/evoked	Spontaneous/evoked	Spontaneous/evoked (CAP A_1)
Electrophysiology	Enlarged nonspecific evoked potential	Late slow wave component of evoked potentials, cortical down state, limited connection with cognition	Frontal origin (source modeling), traveling wave, up-down state alternation, connection with homeostatic and plastic functions
fMRI	BOLD activation in sensorimotor regions + vision, hearing, touch	Activation of auditory cortex + sensory areas, thalamic activation, DMN activation	Frontal sources

Summing up, within the continuum of reactive sleep graphoelements, K-complex seems to be situated between the evoked potential type and reactive slow wave type responses (Table 7.2). The functional significance of this Janus-faced character is that during K-complex, the brain decides "not to wake up" and compensates the disturbing effect of the incoming stimulus by producing slow wave activity.

7.6 Ontogenetic Aspects of Slow Wave Activity in Sleep

The EEG shows recurrent delta waves joined by 8–25-Hz spindle-like oscillatory bursts, called "delta brushes," during the third trimester of gestation. The phenomenon has been known for long time seen in classical descriptions of the neonatal EEG. Studied in preterm infants, these patterns nowadays have obtained functional significance recently. Khazipov et al. (2004) noticed that hand and foot movements of the human neonates heralded the appearance of delta brushes in the corresponding areas of the cortex. Direct hand and foot stimulation also evoked delta brushes over the same areas.

These results highlighted the role of sensory feedback from spontaneous fetal movements, triggering delta-brush oscillations in consistent somatotopic regions of the central cortex. They assumed that before the fetal brain has received elaborated sensory input from the environment, spontaneous movements provide sensory stimulation and drive delta-brush oscillation, contributing to the formation of cortical body maps.

Similar "slow activity transients" like the delta brushes were observed in the rat visual cortex during the second postnatal week. Colonnese and Khazipov (2010) have shown that rhythmic beta oscillations nested by infraslow oscillations sharing several characteristics of delta brushes are driven by certain retinal waves transferred through the lateral geniculate nucleus to cortex. The visual input proved to be transformed to beta oscillations by the thalamocortical system.

Thus, slow activity associated with beta oscillations of rat visual cortex during postnatal weeks appears to be rather the cortical response to visual input than an internally generated spontaneous cortical activity. Taking into account that the synchronized oscillatory network is believed to be essential for the generation of neuronal circuits, both in the visual and somatosensory cortex, the afferent input seems to play an important role in this process.

Here, we see the first example how delta oscillation associated with rapid spindling is the agent of plastic changes of the cortex, both in the neonatal period of rodents and human neonates.

Observations showing that early sensory deprivation in animal life reduces sleep slow wave activity (Miyamoto et al. 2003) seem to be congruent with the role of sensorial input in the development of cortical functions: for example, input-dependent delta activity seems to be an essential element of plastic changes, as early as in the neonatal period of development.

Delta and/or slow EEG activities are of special interest for the studies examining the ontogeny of neuronal systems during childhood and adolescence. A striking parallelism between changes in the amplitude of sleep slow wave activity, cortical metabolism, and synaptic density was evident since the observations of Feinberg et al. (1990). Sleep slow waves (amplitude and/or power spectral density) were shown to increase from birth to the prepubertal ages. Thereafter, a steep decline in sleep slow wave activity is evident in all studies focusing on the age dependency of these measures. The decline was shown to be maximal during adolescence and appeared earlier in girls than in boys. This is congruent with the earlier biological maturation of girls. The spectral power of sleep slow wave activity was shown to correlate positively with the gray matter volume of several cortical areas in children and adolescents between 8 and 19 years of age. The associations were particularly evident in those regions characterized by an age-dependent decrease in gray matter volume (Buchmann et al. 2011). Besides amplitude and/or spectral power of sleep slow wave activity, the steepness of the slopes of sleep slow waves were shown to be different in the age groups: prepubertal children were shown to be characterized by steeper slopes than mature adolescents, both during baseline and after sleep deprivation. This difference is still present after the control of overall amplitude differences (Kurth et al. 2010a). The pattern of anteroposterior differences in sleep slow waves is particularly informative from the point of view of cortical maturation. That is, the frontal predominance of slow wave activity during sleep is gradually achieved during development. The pattern of anteroposterior difference is parallel with the anteroposterior gradient in cortical maturation (Kurth et al. 2010b). The differences between the production and the elimination of cortical synapses were related to the age dependency of sleep slow wave activity. During adolescence, elimination dominates production. This is associated with a decline in slow wave activity during NREM sleep. When the production of cortical synapses is more efficient than their elimination (from birth until the prepubertal age), slow wave activity is high and increasing, and elimination is associated with decreasing amounts of sleep slow wave activity in adults (Ringli and Huber 2011). Moreover, computer simulations of the thalamocortical system suggest that a greater density or greater efficacy of cortical synapses or both are the potential basis of sleep slow waves with steeper slopes. Last, but not the least, aging, and especially pathological aging, is associated with a decrease in efficient synaptic strength and decrease in cells/synapses. This is parallel with the decreased slow wave sleep and slow wave activity in aged and especially pathologically aged humans.

If we consider reactive slow elements separately, the same maturational features emerge. CAP A_1 subtype is preponderant in the peripubertal period, thereafter showing a gradual decrease (Parrino et al. 2012, see Fig. 3.8). An age-dependent decrease in the amplitude of the K-complexes is a similar phenomenon (Colrain 2005).

7.7 Phylogenetic Aspects of Slow Wave Activity

7.7.1 Peculiarities in the EEG Amplitude of Poikilotherms

We have seen that the amplitude of the EEG is mainly determined by its low-frequency components, at least during physiological conditions. This is related to the power-law relationship between EEG amplitude and frequency (Feinberg et al. 1984) or to the exponential decay (of the $1/f^{\alpha}$-type) of spectral power along the frequency axis (Pereda et al. 1998; Freeman et al. 2007). It was repeatedly shown that several non-human or even nonmammalian species possess an ongoing neuronal population activity (we will call it EEG) with similar amplitude-frequency relationship. This was evidently outlined for several mammalian, reptilian, and amphibian species, as well as for fishes (Bullock and Basar 1988). The states of sleep/behavioral rest and waking/behavioral alertness seem to be almost universal in the animal kingdom. Based on the similarities between the amplitude-frequency relationship of the reptilian and mammalian EEG, we would expect larger EEG amplitudes in sleeping or resting reptiles when compared to the behaviorally active ones. At least this is what we usually see in mammals. However, this is not the case in poikilotherm vertebrates, namely, fishes, amphibians, and reptiles. For example, the EEG power of the lizard Gallotia galloti was shown to be higher during active waking than during passive waking. Both were higher than the EEG power of the state of nocturnal rest (Fig. 7.3). As regards sensory stimulation, post-stimulation amplitudes of the EEG significantly exceeded the amplitude of brain electrical activity recorded during the pre-stimulation period (Nicolau et al. 2000; Rial et al. 2010; Fig. 7.4). The above results seem to be a general feature of the brain physiology of reptiles, amphibians, and fishes. This thesis is supported by 14 early reports, published between 1964 and 1988, all highlighting higher amplitude of brain electrical activity during states of alert waking and a lower one during rest (Nicolau et al. 2000). Thus, unlike in mammals, states of higher activity and arousal are associated with higher amplitude of brain electrical activity in poikilotherm vertebrates. The higher EEG amplitude during/after stimulation or behavioral arousal is reminiscent for the reactive delta activity or K-complexes during mammalian slow wave sleep. But, are these high-amplitude waves slow waves or some other frequency components of the EEG?

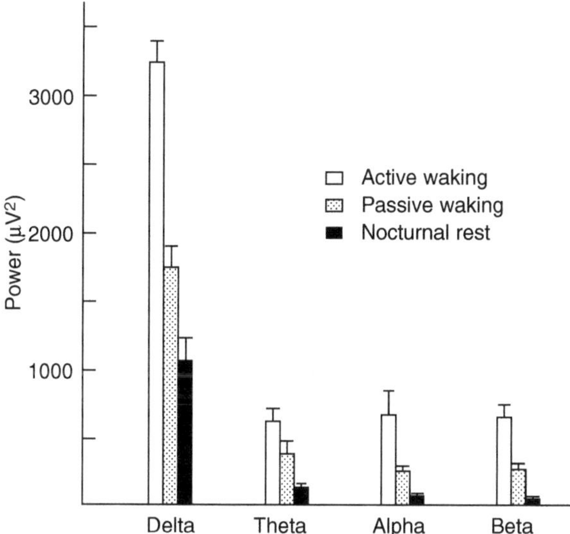

Fig. 7.3 Total absolute power in delta (1–4 Hz), theta (4–8 Hz), alpha (8–13 Hz), and beta (13–30 Hz) EEG bands as recorded in day active, day resting, and night resting *Gallotia galloti* lizards (By the courtesy of Ruben Rial Planas)

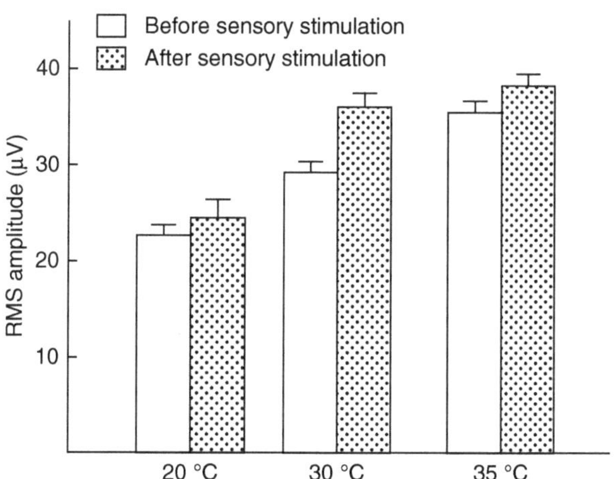

Fig. 7.4 Effects of the sensory stimulation in the root mean square (RMS) amplitude of the EEG in *Gallotia galloti* lizards as recorded in hypothermic (20 °C), euthermic (30 °C), and hyperthermic (35 °C) environments (By the courtesy of Ruben Rial Planas)

7.7.2 Peculiarities in the Slow Waves of Poikilotherms: Vigilance Level and Responsivity

The direct recordings from the cortical surface of the lizard Gallotia galloti were shown to contain well-formed slow waves and groups of waves with spindle-like morphology (De Vera et al. 1994; Nicolau et al. 2000). Just as the general amplitude of brain electrical activity, these slow waves were shown to be characterized by higher amplitude during periods of behavioral and sensory arousal (Fig. 7.5) and by higher amplitude during higher ambient temperatures in the poikilotherm vertebrates. Again, it is important to note that high body temperatures induced by passive

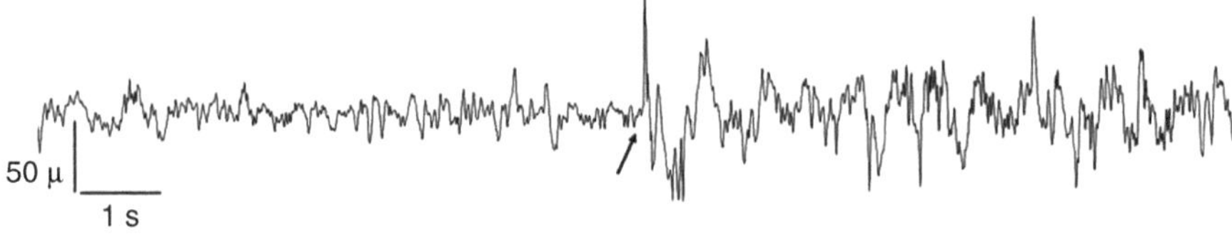

Fig. 7.5 Sample of EEG obtained from a *Gallotia galloti* lizard at 30 °C, showing the effects of a tap in the walls of the recording chamber (*arrow*) (By the courtesy of Ruben Rial Planas)

Table 7.3 Similarities of the evoked and spontaneous slow waves of poikilotherm vertebrates with the human K-complexes

	Evoked and spontaneous slow potentials in poikilotherm vertebrates	K-complexes in humans
Dominant frequency	Species-specific but always slow	Slow (0.5–2 Hz)
Stimulation dependence	Spontaneous or stimulation-induced	Spontaneous or stimulation-induced
Wave morphology	Polyphasic in some cases initiated by a sharp component	Biphasic in some cases initiated by a sharp component
Modality of the inducing stimuli	Each studied modality	Each studied modality
Post-potential background activity changes	Characteristic, increases in high frequencies	Characteristic and well specified (K-sigma, K-alpha, K-delta complexes, etc.)

heating or physical activity can induce significant increases in sleep slow wave activity of humans. That is, high temperatures increase slow wave activity during behaviorally active periods of reptiles and during slow wave sleep of mammals.

Reported slow waves in the lizard Gallotia galloti are of various frequencies. A low-frequency spectral peak between 0.5 and 4 Hz is present at different body temperatures (De Vera et al. 1994). Nicolau et al. (2000) reported occasional slow waves in the posterior cortex mainly, which were shown to be independent of breathing and cortical or olfactory spindles. They also presented evidences for well-formed slow waves appearing during continuous sensory stimulation, that is, the effects of lights. In the paper of Rial et al. (2010), the same group presented continuous slow wave activity mixed with high-frequency waves in behaviorally active reptiles. Based on the figures and reported frequency bands of several papers, it seems that these slow waves cover the slow oscillation, as well as the delta and theta waves. It has to be noted, however, that there is no unequivocal evidence for the categorization of different types of slow waves in poikilotherm vertebrates (Table 7.3).

Indirect evidence for the specific dynamism and reactivity of reptilian slow wave activity comes from the studies showing an increased relative predominance of 0.3–7.5-Hz band power of the Gallotia galloti lizards' brain electrical activity during states of eyes open compared to the states with eyes closed. Moreover, a specific interaction between temperature and eyes open/closed states was shown to shape the slow waves of the same species. Namely, a temperature-dependent increase of harmonic power is invariantly observed in the higher frequencies (8–30 Hz). However, the harmonic power of the slower waves (0.3–7.5 Hz) is shaped by both the ambient temperature and sensory arousal: these waves are temperature-dependently increased in the states with open eyes (sensory stimulation), but not in the states with eyes closed. Thus, the slow waves are the ones reflecting more directly and specifically the sensory/behavioral arousal. Based on the above findings, it is tempting to say that slow wave activity is reactive in nature being an activation-related phenomenon in reptiles.

Last, but not the least, there are reported similarities between reptilian brain electrical activity and human slow wave sleep EEG. These similarities are evident for the Gallotia galloti lizard at 25 grades of Celsius and are defined by a main spectral peak with a central frequency of ~1 Hz, with the presence of nonlinear structure and the lack of correlation between the correlation dimension and the fractal exponent derived from coarse graining spectral analysis.

It is important to note that some influential studies deny the existence of slow waves in reptiles and other species lacking a well-developed pallium (Rattenborg 2006). This is an especially controversial issue because the lack of some kind of brain electrical activity cannot be concluded if one uses the test of a visual inspection only. For a mammalian example we could invoke the case of slow waves during wakefulness in humans and monkeys. It is clear the slow waves are not particularly striking components of the mammalian EEG during wakefulness, yet these relatively low-amplitude waves were shown to participate in attention and stimulus detection (Lakatos et al. 2008). As we have seen, the $1/f^{\alpha}$-type spectra of reptilian brain electrical activity imply the existence of some kind of slow activity. The amplitude of this activity is clearly lower than the amplitude of the slow waves in mammalian and avian sleep. But, the question of the amplitude of ongoing brain electrical activity is class- and species-specific and is related to the level of telencephalization; hence, the differences in amplitude are rather a natural consequence of brain evolution. The question is whether the specific dynamic nature of this slow wave activity reflects some specific rules, which were uncovered in other species – especially mammalian ones – before.

Effects of external stimuli are usually studied by the evoked potential technique in mammals. This implies the use of several repetitions of the same stimuli and the averaging of the post-stimulation EEG records in order to filter out the so-called noise and retain the stimuli-dependent activities. It is worth noting that such techniques are seldom used in nonmammalian species. This could be caused by the strikingly visible effects of external stimuli on the ongoing brain electrical activity of poikilotherms. A common feature of this evoked activity is a large-amplitude slow component and a change in frequency of the background activity that follows it. Earlier reports are common in not even mentioning the former and focusing on the latter aspect only.

External stimuli seem to interrupt the ongoing brain electrical activity of lizards Gekko gecko by high-amplitude, sharp-slow complexes, which remain uncommented, but clearly visible in the figures reported by Gaztelu et al. (1991). These complexes are followed by an increase in faster activities thereafter. Similar events are observed in the lizard Sceloporus olivaceus. An air puff may induce a large sharp-slow complex followed by a change in background activity (Hunsaker and Lansing 1962). Weak electrical stimulation of elephantnose fishes (Gnathonemus petersii) induces a large-amplitude slow component termed delta F wave by Prechtl et al. (1998). Delta refers to the change and not to the slow nature of these potentials here: that is, a change in frequency (fastening) follows these potentials. But the potentials are of large amplitude and slow indeed, thus the term delta derived from the EEG nomenclature could be coherent with the nature of these waves too. The appearance of a stripe (moving visual stimuli) induces oscillatory events in the turtle Pseudemys scripta, which are superimposed on a slow potential (Bullock et al. 1993). Last, but not the least, flash evoked potentials were characterized by slow triphasic components with superimposed spindles in Gallotia galloti lizards. These wave complexes were similar to the K-complexes of human sleep (De Vera et al. 1994).

The large slow waves following external stimuli are characteristic features of mammalian slow wave sleep. Humans were shown to produce K-complexes after their sensory stimulation during NREM sleep (Loomis et al. 1938). The slow component of the K-complex seems to be the EEG outcome of an isolated cortical down state (Cash et al. 2009). These isolated down states are preceded in some cases by a localized sharp component corresponding to a stimulation-induced localized up state. Spontaneous K-complexes appear with a frequency of around 1/35 s during NREM sleep (Halász 2005). These K-complexes are similar to the spontaneous slow potentials appearing in the ongoing brain electrical activity of different poikilotherm vertebrate species.

7.7.3 *Peculiarities in the Chemical Background of Arousal in Poikilotherms*

Neuropharmacological studies undoubtedly revealed the relationship between cortical acetylcholine (Ach) and brain electrical potentials. Ach seems to be the most powerful inducer of high-frequency activity in mammalian brain (Jones 2005). The direct relationship between Ach and the suppression of slow wave activity was evidenced in the studies showing that unihemispheric slow waves did negatively correlated with Ach levels in the ipsilateral hemisphere of marine mammals. There was no relationship between unihemispheric slow wave activity and contralateral Ach level. Moreover, the muscarinic antagonist atropine is known to induce slow waves in mammals (Dringenberg and Vanderwolf 1998), including humans (Czopf et al. 1977), whereas during atropine-induced coma, the evoked potentials consist of large-amplitude, biphasic slow wave reminiscent of K-complexes. The amplitude of these evoked slow waves is positively correlated with the deepness of the coma (Czopf et al. 1977).

A major change in the views around the issue of the phylogeny of sleep-waking states was provoked by the reconsideration of the physiology of the platypus and of other monotremes. These primitive mammals were thought to miss the state of REM sleep. However, later studies revealed that these animals are indeed champions of REM sleep time, but that these REM sleep periods are characterized by slow cortical waves instead of the well-known fast activity seen in other mammals. The lack of a significant decrease in slow waves during full-blown REM states was attributed to peculiarities of the cholinergic system of the platypus. The cholinergic nuclei of the hypothalamus were shown to be missing in this species, which might contribute to a decreased cortical cholinergic tone during REM sleep. It is unclear whether the lack of an efficient Ach system is related to the lack of the classical arousal reaction in the poikilotherm vertebrates. Earlier reports suggested the lack of cholinergic interneurons in the reptilian brain. This could be suggestive, but not a very strong evidence yet. As we previously mentioned, direct effects of atropine on mammals are known to be characterized by the increase in slow wave activity. Such effects of atropine are lacking in the lizard Gallotia galloti. Thus, muscarinic receptors are not participating in the process of initiation/suppression of slow waves in these reptiles. Other reported pharmacological probes include the α1-adrenergic antagonist prazosin and the β-blocker propranolol. Both reduced significantly the slow wave activity of lizards. Moreover, both prazosin and propranolol were shown to reduce slow wave sleep or slow wave activity in mammals (Hilakivi et al. 1978, 1987). Thus, the adrenergic system seems to play similar roles in shaping the slow wave activity of reptiles and mammals.

There are other endogenous neurochemical agents characterized by a prominent role in sleep-wake and arousal regulations. The neuropeptide orexin localized in the lateral hypothalamus was shown to be released during wakefulness, stimulate the major arousal systems of the brain, and suppress the activity of the hypothalamic VLPO GABA-ergic and galaninergic sleep-inducing neurons. Orexin was shown to be present in the brain of several nonmammalian species, but there is evidence for the fact that unlike in mammals, orexinergic neurons of the goldfish (*Carassius auratus*) do not form regular synapses with neurons of the arousal-inducing aminergic nuclei (Huesa et al. 2005). Although this view is not shared by many comparative neuroanatomists, who present data on the close vicinity of orexinergic fibers with noradrenergic and serotoninergic cells in fishes and lizards (Panula 2010; Domínguez et al. 2010), neither of these authors specifically addresses the question put by Huesa et al. (2005). Do orexinergic fibers form synapses with the somata or proximal dendrites of the aminergic cells? If the answer to the above question is no, or is just less than in mammals, this could be suggestive for the lack of the flip-flop circuitry-stabilizing arousal in these nonmammalian species. Such a lack may result in the lack of the long-term homeostasis of slow waves which is well documented in mammals (Borbély 2001). The short-term dynamical regulation or the instant homeostasis is present indeed. Stimuli induce a parallel increase in low- and high- frequency activity, which we hypothesize to be a kind of balanced cortical reaction providing the stabilization of the network by a so-called delta injection.

The above hypothesis is congruent with the view that reptilian wakefulness is some kind of NREM sleep-like state. According to this view mammalian wakefulness is a phylogenetically new behavioral state, which has no antecedents in the pre-mammalian species. The old type of wakefulness, which was characterized by the reactive slow wave type of functioning mode, was indeed suppressed, its daily emergence being limited to the periods of behavioral rest/sleep (Nicolau et al. 2000; Rial et al. 2010). This sequence of phylogenetic adaptation could lead to the appearance of the long-term homeostasis of slow waves (accumulation of the need for slow waves during wakefulness and their recuperation during sleep), which are indeed largely suppressed during the new mammalian type of wakefulness. There are at least two additional issues supporting this hypothesis. One is the parallelism between the positive neuroanatomical correlates of sleep slow waves in the human brain (larger brain associated with higher sleep slow wave activity) and a similar correlation between telencephalization and EEG amplitude of different species. The latter is evidenced by several sources and scientific reports. First, it was Bullock and Basar (1988), who outlined that the EEG amplitude is increasing in the fish-amphibian-reptile series. These classes form an unquestionable series in telencephalization (fishes<amphibians<reptiles), which seems to be parallel with their increasing EEG amplitude. Secondly, there is a characteristic species-specific increase in sleep EEG amplitude of several mammalian species directly compared by Kleinlogel (1983). The species investigated so far were the human, the rat, the Syrian hamster (Mesocricetus auratus), the guinea pig (Cavia porcellus), and the Vietnamese potbellied pig (Sus scrofa f. domestica) (Kleinlogel 1983). The largest slow waves (relative to the EEG of the wakeful state) were recorded in humans. The precise order was the following: human>Vietnamese pig>rat>guinea pig>hamster. This result is coherent with the view that the slow waves are parts of some reorganization in large-scale neuronal assemblies which are especially needed in beings with high levels of telencephalization.

The predominance of reactive slow waves in the EEG of poikilotherm vertebrates and the peculiar parallelism between the dynamics of mammalian sleep slow waves and reptilian activity-dependent slow waves are changing the views on the ontogeny of sleep. The delta brushes presented before are in fact activation-induced reactive slow wave groups in early (preterm) infancy which are congruent with the Haeckel's law claiming that ontogeny repeats phylogeny. Similar signs for this kind of pre-mammalian physiology are the non-separability of NREM and REM sleep in early infancy as well as the temporal coincidence of eye movements and slow waves during sleep.

7.8 The Slow Waves in REM Sleep

Although slow waves are considered as hallmarks of NREM sleep, some forms of slow wave EEG activity are measurable in REM sleep too. It is worth noting that the level of objectively measured slow wave activity in REM sleep is usually intermediate between NREM sleep and wakefulness (NREM>REM>wakefulness). Theories are very restrictive in explaining the nature and origin of slow wave activity in REM sleep. Here we provide some clues aiming to focus the attention of the scientific community on some of these potentially important aspects of REM sleep.

The slow wave activity in the human EEG has state-unspecific aspects, as increases in frontal 1–7-Hz power after sleep deprivation were evidenced for both NREM and REM phases of recovery sleep. Moreover, the sleep deprivation-induced increases in NREM and REM sleep 2–7-Hz activity correlated positively across the subjects (Marzano et al. 2010). Similar observations were made for wakefulness, but for this state, the sleep deprivation-induced increase was largest in (but not limited to) the theta band (Tinguely et al. 2006). Thus, the long-term homeostasis of the slow waves is evidently present in all three behavioral states. However, we have no evidence for the instant delta homeostasis or delta injection (reactive delta) in REM sleep or wakefulness. That is, the bursting mode of the thalamocortical system is a necessary prerequisite of reactive slow wave EEG activity, while the slowing of the background EEG can occur in the tonic firing mode also. The phenomenon of local sleep could be a physiological framework of this latter case.

Hippocampal theta waves or hippocampal rhythmic slow wave activity was shown to be a main characteristic of explorative behavior and REM sleep in many mammalian species (Grastyan and Karmos 1961; Vanderwolf 1969; Arnolds et al. 1979; Robinson 1980). When aiming to describe and characterize this phenomenon during human REM sleep, we relied on foramen ovale records of epileptic patients undergoing presurgical evaluation. These electrodes are inserted in the ambient cistern, are parallel with the long axis of the hippocampus, and record parahippocampal activity. We recorded rhythmic waves of 1.5–3-Hz frequency specifically associated with the states of REM sleep (Bódizs et al. 2001; Fig. 7.6). The waves were continuous during the whole REM periods, and no other frequency had this high rhythmicity. We concluded that this oscillation is the counterpart of the hippocampal theta of mammalian REM sleep and that the 1.5–3-Hz delta EEG activity is a basic neurophysiologic feature of human REM sleep. The REM sleep delta waves recorded in this clinical situation were grouping high-frequency activities, including gamma activity, just as it was observed in the case of hippocampal theta waves in rodents (Clemens et al. 2009). Further evidence for REM sleep-related rhythmic hippocampal delta activity (2–4 Hz) of human subjects was presented by Moroni et al. (2011) in a recent study. Given these evidences, it is tempting to say that the hippocampal RSA of humans is in fact a delta frequency activity, contributing to the temporal and perhaps posterior delta power measured during REM sleep (Bódizs et al. 2001, 2005). Moreover, these findings fit well with the allometric theory of the frequency of hippocampal RSA, suggesting that larger brains display slower hippocampal RSA (Blumberg 1989). The phase coupling of hippocampal delta with gamma in human REM sleep is reminiscent for the phase coupling between the slow oscillation and fast EEG activities during NREM sleep (Steriade 2006). Although we have no data on this issue, given the nature of the hippocampal RSA, it is improbable that REM sleep hippocampal delta activity is homeostatically regulated, and we do not know whether it could be triggered or evoked by external stimuli.

Another source of (posterior) slow wave activity in human REM sleep was reported by Magnin et al. (2004). The authors recorded thalamic activity in epileptic patients implanted with intracranial electrodes and found a prominent delta (1.5–4 Hz) activity in the medial pulvinar nucleus during REM sleep. This was

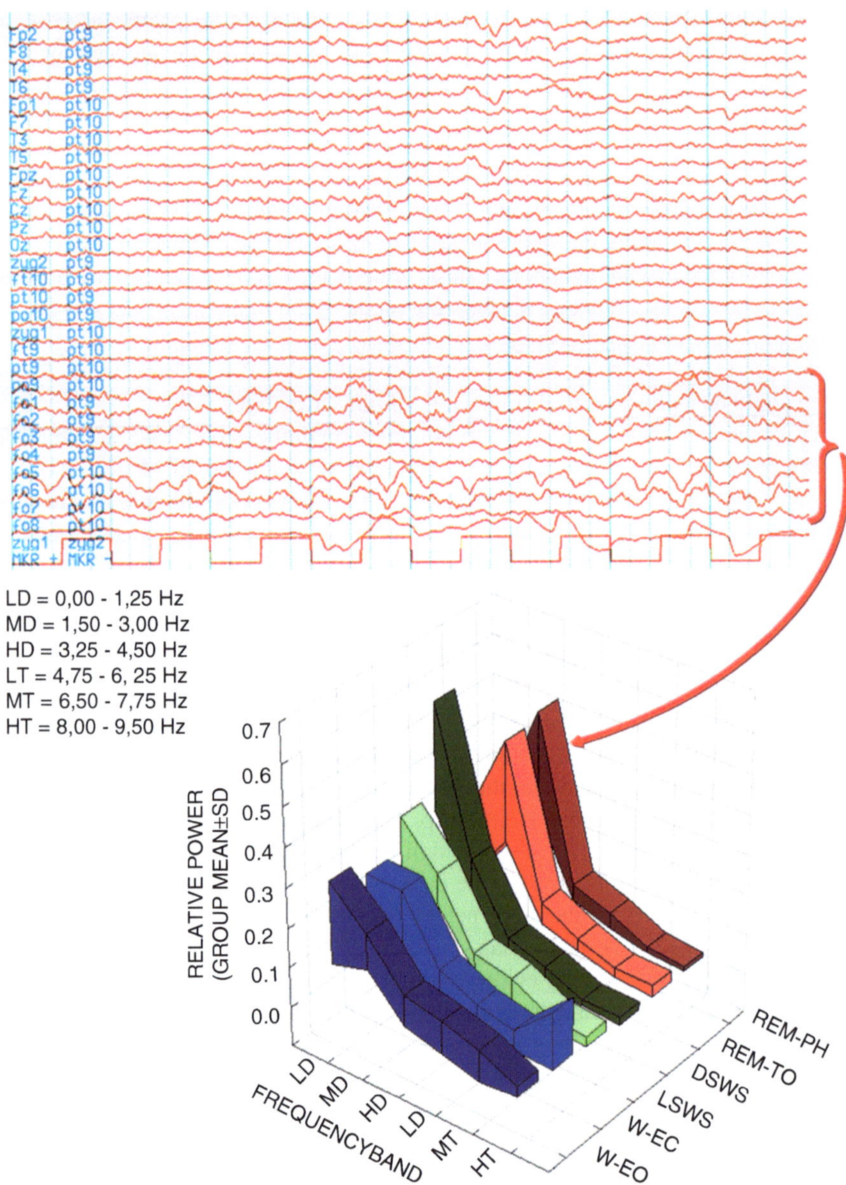

Fig. 7.6 Characteristic REM sleep delta (~2 Hz) activity near the parahippocampal gyrus in an adult human subject implanted with foramen ovale electrodes (fo1–8). Insertion is a part of the presurgical examination of epilepsy. *Top*: Scalp EEG shows low-amplitude EEG activity typical for REM sleep, with occasional eye movement artifacts. The contralateral mastoid is used as EEG reference. Calibration marks: 100 μV and 1 s. *Bottom*: State dependency of the relative power of foramen ovale recorded EEG in wakefulness-eyes open (*W-EO*), wakefulness-eyes closed (*W-EC*), light slow wave sleep (*LSWS*), deep slow wave sleep (*DSWS*), REM sleep tonic phases (*REM-TO*), and REM sleep phasic phases (*REM-PH*). Frequency bands are the following: *LD* (low delta), *MD* (mid-delta), *HD* (high delta), *LT* (low theta), (*MT*) mid theta, (*HT*) and high theta

interpreted as a lack of activation of the human medial pulvinar nucleus during REM sleep hypothesized to contribute to the cognitive peculiarities of dreams. Again, we have no data for the homeostasis of REM sleep medial pulvinar delta activity or on its reactivity.

The above findings suggest that there are slow waves with specific origin and function being characteristic for REM instead of NREM sleep. Although their detailed investigation waits for further research, it seems that they are of different sources and have different functions compared to the NREM sleep slow waves.

References

Achermann P, Borbély AA. Low-frequency (<1 Hz) oscillations in the human sleep electroencephalogram. Neuroscience. 1997;81(1):213–22.

Albrecht D, Royl G, Kaneoke Y. Very slow oscillatory activities in lateral geniculate neurons of freely moving and anesthetized rats. Neurosci Res. 1998;32(3):209–20.

Amzica F, Steriade M. Cellular substrates and laminar profile of sleep K-complex. Neuroscience. 1998;82(3):671–86.

Anderson C, Horne JA. Prefrontal cortex: links between low frequency delta EEG in sleep and neuropsychological performance in healthy, older people. Psychophysiology. 2003;40(3):349–57.

Arnolds DE, Aitink JW, Boeyinga P, Lopes da Silva FH. Hippocampal EEG in dog, cat and man (author's transl). Rev Electroencephalogr Neurophysiol Clin. 1979;9(4):326–32.

Aserinsky E, Kleitman N. Regularly occurring periods of eye motility, and concomitant phenomena, during sleep. Science. 1953;118(3062):273–4.

Bersagliere A, Achermann P. Slow oscillations in human non-rapid eye movement sleep electroencephalogram: effects of increased sleep pressure. J Sleep Res. 2010;19(1 Pt 2):228–37.

Blumberg MS. An allometric analysis of the frequency of hippocampal theta: the significance of brain metabolic rate. Brain Behav Evol. 1989;34(6):351–6.

Bódizs R, Kántor S, Szabó G, Sz cs A, Er ss L, Halász P. Rhythmic hippocampal slow oscillation characterizes REM sleep in humans. Hippocampus. 2001;11:747–53.

Bódizs R, Békésy M, Sz cs A, Barsi P, Halász P. Sleep-dependent hippocampal slow activity correlates with waking memory performance in humans. Neurobiol Learn Mem. 2002;78(2):441–57.

Borbély AA, Baumann F, Brandeis D, Strauch I, Lehmann D. Sleep deprivation: effect on sleep stages and EEG power density in man. Electroencephalogr Clin Neurophysiol. 1981;51(5):483–95.

Braun AR, Balkin TJ, Wesenten NJ, Carson RE, Varga M, Baldwin P, Selbie S, Belenky G, Herscovitch P. Regional cerebral blood flow throughout the sleep-wake cycle. An H2(15)O PET study. Brain. 1997;120(Pt 7):1173–97.

Buchmann A, Kurth S, Ringli M, Geiger A, Jenni OG, Huber R. Anatomical markers of sleep slow wave activity derived from structural magnetic resonance images. J Sleep Res. 2011;20(4):506–13.

Bullock TH, Basar E. Comparison of ongoing compound field potentials in the brains of invertebrates and vertebrates. Brain Res. 1988;472(1):57–75.

Bullock, TH and Achimowicz JZ. A comparative survey of oscillatory brain activity, especially gamma-band rhythms. In: Symposium on "oscillatory event related brain dynamics", Tecklenburg/Münsterland, 1993.

Cirelli C, Huber R, Gopalakrishnan A, Southard TL, Tononi G. Locus ceruleus control of slow-wave homeostasis. J Neurosci. 2005;25(18):4503–11.

Clemens Z, Weiss B, Szucs A, Eross L, Rásonyi G, Halász P. Phase coupling between rhythmic slow activity and gamma characterizes mesiotemporal rapid-eye-movement sleep in humans. Neuroscience. 2009;163(1):388–96.

Colonnese MT, Khazipov R. “Slow activity transients” in infant rat visual cortex: a spreading synchronous oscillation patterned by retinal waves. J Neurosci. 2010;30(12):4325–37.

Csercsa R, Dombovári B, Fabó D, Wittner L, Eross L, Entz L, Sólyom A, Rásonyi G, Szucs A, Kelemen A, Jakus R, Juhos V, Grand L, Magony A, Halász P, Freund TF, Maglóczky Z, Cash SS, Papp L, Karmos G, Halgren E, Ulbert I. Laminar analysis of slow wave activity in humans. Brain. 2010;133(9):2814–29.

Czisch M, Wetter TC, Kaufmann C, Pollmächer T, Holsboer F, Auer DP. Altered processing of acoustic stimuli during sleep: reduced auditory activation and visual deactivation detected by a combined fMRI/EEG study. Neuroimage. 2002;16(1):251–8.

Czisch M, Wehrle R, Kaufmann C, Wetter TC, Holsboer F, Pollmächer T, Auer DP. Functional MRI during sleep: BOLD signal decreases and their electrophysiological correlates. Eur J Neurosci. 2004;20(2):566–74.

Czisch M, Wehrle R, Stiegler A, Peters H, Andrade K, Holsboer F, Sämann PG. Acoustic oddball during NREM sleep: a combined EEG/fMRI study. PLoS One. 2009;4(8):e6749.

Czopf J, Karmos G, Gaszner P, Kelényi L. A vizuális kiváltott válasz változása terápiás atropin coma alatt. Ideggyogy Sz. 1977;30:81–9.

Dang-Vu TT, Schabus M, Desseilles M, Albouy G, Boly M, Darsaud A, Gais S, Rauchs G, Sterpenich V, Vandewalle G, Carrier J, Moonen G, Balteau E, Degueldre C, Luxen A, Phillips C, Maquet P. Spontaneous neural activity during human slow wave sleep. Proc Natl Acad Sci USA. 2008;105(39):15160–5.

De Vera L, González J, Rial RV. Reptilian waking EEG: slow waves, spindles and evoked potentials. Electroencephalogr Clin Neurophysiol. 1994;90(4):298–303.

Depootere H, Granger P, Leonardon J, Terzano MG. Evaluation of cyclic alternating pattern in rats by automatic analysis of sleep amplitude variations. Effect of zolpidem. In: Terzano MG, Halász P, Declerck AC, editors. Phasic events and dynamic organization of sleep. New York: Raven Press; 1991. p. 17–33.

Domínguez L, Morona R, Joven A, González A, López JM. Immunohistochemical localization of orexins (hypocretins) in the brain of reptiles and its relation to monoaminergic systems. J Chem Neuroanat. 2010;39(1):20–34.

Dringenberg HC, Vanderwolf CH. Involvement of direct and indirect pathways in electrocorticographic activation. Neurosci Biobehav Rev. 1998;22(2):243–57.

Ehlers CL, Foote SL. Ultradian periodicities in EEG and behavior in the squirrel monkey (*Saimiri sciureus*). Am J Primatol. 1984;7:381–9.

Feinberg I, Thode Jr HC, Chugani HT, March JD. Gamma distribution model describes maturational curves for delta wave amplitude, cortical metabolic rate and synaptic density. J Theor Biol. 1990;142(2):149–61.

Fischgold H, Mathis P. Obnubilations, comas et stupeurs. Etudes electroenceph. Paris: Masson et Cie; 1959.

Gaztelu JM, García-Austt E, Bullock TH. Electrocorticograms of hippocampal and dorsal cortex of two reptiles: comparison with possible mammalian homologs. Brain Behav Evol. 1991;37(3):144–60.

Grastyan E, Karmos G. A study of a possible “dreaming” mechanism in the cat. Acta Physiol Acad Sci Hung. 1961;20:41–50.

Halász P. K-complex, a reactive EEG graphoelement of NREM sleep: an old chap in a new garment. Sleep Med Rev. 2005;9(5):391–412.

Halász P. The role of the non-specific sensory activation in sleep regulation and in the pathomechanism of generalized epilepsy with generalized spike wave discharge. Doctoral thesis, The Hungarian Academy of Sciences, Budapest; 1982.

Hanlon EC, Vyazovskiy VV, Faraguna U, Tononi G, Cirelli C. Synaptic potentiation and sleep need: clues from molecular and electrophysiological studies. Curr Top Med Chem. 2011;11(19):2472–82.

Hilakivi I, Mäkelä J, Leppävuori A, Putkonen PT. Effects of two adrenergic beta-receptor blockers on the sleep cycle of the cat. Med Biol. 1978;56(3):138–43.

Hilakivi I, Kovala T, Leppävuori A, Shvaloff A. Effects of serotonin and noradrenaline uptake blockers on wakefulness and sleep in cats. Pharmacol Toxicol. 1987;60(3):161–6.

Hofle N, Paus T, Reutens D, Fiset P, Gotman J, Evans AC, Jones BE. Regional cerebral blood flow changes as a function of delta and spindle activity during slow wave sleep in humans. J Neurosci. 1997;17(12):4800–8.

Huesa G, van den Pol AN, Finger TE. Differential distribution of hypocretin (orexin) and melanin-concentrating hormone in the goldfish brain. J Comp Neurol. 2005;488(4):476–91.

Hughes SW, Lorincz ML, Parri HR, Crunelli V. Infraslow (<0.1 Hz) oscillations in thalamic relay nuclei basic mechanisms and significance to health and disease states. Prog Brain Res. 2011;193:145–62.

Hunsaker 2nd D, Lansing RW. Electroencephalographic studies of reptiles. J Exp Zool. 1962;149:21–32.

Jahnke K, von Wegner F, Morzelewski A, Borisov S, Maischein M, Steinmetz H, Laufs H. To wake or not to wake? The two-sided nature of the human K-complex. Neuroimage. 2012;59(2): 1631–8.

Jones BE. From waking to sleeping: neuronal and chemical substrates. Trends Pharmacol Sci. 2005;26(11):578–86.

Jouvet M, Michel F, Courjon J. On a stage of rapid cerebral electrical activity in the course of physiological sleep. C R Seances Soc Biol Fil. 1959;153:1024–8.

Kaufmann C, Wehrle R, Wetter TC, Holsboer F, Auer DP, Pollmächer T, Czisch M. Brain activation and hypothalamic functional connectivity during human non-rapid eye movement sleep: an EEG/fMRI study. Brain. 2006;129(Pt 3):655–67.

Khazipov R, Sirota A, Leinekugel X, Holmes GL, Ben-Ari Y, Buzsáki G. Early motor activity drives spindle bursts in the developing somatosensory cortex. Nature. 2004;432(7018): 758–61.

Kleinlogel H. Sleep in various species of laboratory animals. Neuropsychobiology. 1983;9(2–3): 174–7.

Kurth S, Jenni OG, Riedner BA, Tononi G, Carskadon MA, Huber R. Characteristics of sleep slow waves in children and adolescents. Sleep. 2010a;33(4):475–80.

Kurth S, Ringli M, Geiger A, LeBourgeois M, Jenni OG, Huber R. Mapping of cortical activity in the first two decades of life: a high-density sleep electroencephalogram study. J Neurosci. 2010b;30(40):13211–9.

Lakatos P, Karmos G, Mehta AD, Ulbert I, Schroeder CE. Entrainment of neuronal oscillations as a mechanism of attentional selection. Science. 2008;320(5872):110–3.

Lambertz M, Langhorst P. Simultaneous changes of rhythmic organization in brainstem neurons, respiration, cardiovascular system and EEG between 0.05 Hz and 0.5 Hz. J Auton Nerv Syst. 1998;68(1–2):58–77.

Lee J, Kim D, Shin HS. Lack of delta waves and sleep disturbances during non-rapid eye movement sleep in mice lacking alpha1G-subunit of T-type calcium channels. Proc Natl Acad Sci USA. 2004;101(52):18195–9.

Loomis AL, Harvey EN, Hobart G. Further observations on the potential rhythms of the cerebral cortex during sleep. Science. 1935;82(2122):198–200.

Loomis AL, Harvey EN, Hobart GA. Distribution of disturbance-patterns in the human electroen cephalogram, with special reference to sleep. J Neurophysiol. 1938;1(5):413–30.

Lörincz ML, Geall F, Bao Y, Crunelli V, Hughes SW. ATP-dependent infra-slow (<0.1 Hz) oscillations in thalamic networks. PLoS One. 2009;4(2):e4447.

Magnin M, Bastuji H, Garcia-Larrea L, Mauguière F. Human thalamic medial pulvinar nucleus is not activated during paradoxical sleep. Cereb Cortex. 2004;14(8):858–62.

Maquet P. Functional neuroimaging of normal human sleep by positron emission tomography. J Sleep Res. 2000;9(3):207–31.

Maquet P, Phillips C. Functional brain imaging of human sleep. J Sleep Res. 1998;7 Suppl 1:42–7.

Marshall L, Helgadóttir H, Mölle M, Born J. Boosting slow oscillations during sleep potentiates memory. Nature. 2006;444(7119):610–3.

Marzano C, Ferrara M, Curcio G, De Gennaro L. The effects of sleep deprivation in humans: topographical electroencephalogram changes in non-rapid eye movement (NREM) sleep versus REM sleep. J Sleep Res. 2010;19(2):260–8.

Massimini M, Rosanova M, Mariotti M. EEG slow (approximately 1 Hz) waves are associated with nonstationarity of thalamo-cortical sensory processing in the sleeping human. J Neurophysiol. 2003;89(3):1205–13.

McKeown MJ, Humphries C, Achermann P, Borbély AA, Sejnowski TJ. A new method for detecting state changes in the EEG: exploratory application to sleep data. J Sleep Res. 1998;7: 48–56.

Miyamoto H, Katagiri H, Hensch T. Experience-dependent slow-wave sleep development. Nat Neurosci. 2003;6(6):553–4.

Mölle M, Marshall L, Gais S, Born J. Learning increases human electroencephalographic coherence during subsequent slow sleep oscillations. Proc Natl Acad Sci USA. 2004;101(38): 13963–8.

Monto S, Palva S, Voipio J, Palva JM. Very slow EEG fluctuations predict the dynamics of stimulus detection and oscillation amplitudes in humans. J Neurosci. 2008;28(33):8268–72.

Moroni F, Nobili L, De Carli F, Massimini M, Francione S, Marzano C, Proserpio P, Cipolli C, De Gennaro L, Ferrara M. Slow EEG rhythms and inter-hemispheric synchronization across sleep and wakefulness in the human hippocampus. Neuroimage. 2011;60(1):497–504.

Moruzzi G, Magoun HW. Brain stem reticular formation and activation of the EEG. Electroencephalogr Clin Neurophysiol. 1949;1(4):455–73.

Murphy M, Riedner BA, Huber R, Massimini M, Ferrarelli F, Tononi G. Source modeling sleep slow waves. Proc Natl Acad Sci USA. 2009;106(5):1608–13.

Nicolau MC, Akaârir M, Gamundí A, González J, Rial RV. Why we sleep: the evolutionary pathway to the mammalian sleep. Prog Neurobiol. 2000;62(4):379–406.

Nofzinger EA, Mintun MA, Wiseman M, Kupfer DJ, Moore RY. Forebrain activation in REM sleep: an FDG PET study. Brain Res. 1997;770(1–2):192–201.

Novak P, Lepicovska V. Slow modulation of EEG. Neuroreport. 1992;3(2):189–92.

Panula P. Hypocretin/orexin in fish physiology with emphasis on zebrafish. Acta Physiol (Oxf). 2010;198(3):381–6.

Pappenheimer JR, Koski G, Fencl V, Karnovsky ML, Krueger J. Extraction of sleep-promoting factor S from cerebrospinal fluid and from brains of sleep-deprived animals. J Neurophysiol. 1975;38(6):1299–311.

Penttonen M, Nurminen N, Mietinnen R, Sirviö J, Henze DA, Csicsvári J, Buzsáki G. Ultra-slow oscillation (0.025 Hz) triggers hippocampal afterdischarges in Wistar rats. Neuroscience. 1999;94:735–43.

Pereda E, Gamundi A, Rial R, González J. Non-linear behaviour of human EEG: fractal exponent versus correlation dimension in awake and sleep stages. Neurosci Lett. 1998;250(2):91–4.

Picchioni D, Horovitz SG, Fukunaga M, Carr WS, Meltzer JA, Balkin TJ, Duyn JH, Braun AR. Infraslow EEG oscillations organize large-scale cortical-subcortical interactions during sleep: a combined EEG/fMRI study. Brain Res. 2011;1374:63–72.

Prechtl JC, von der Emde G, Wolfart J, Karamürsel S, Akoev GN, Andrianov YN, Bullock TH. Sensory processing in the pallium of a mormyrid fish. J Neurosci. 1998;18(18):7381–93.

Rattenborg NC. Evolution of slow-wave sleep and palliopallial connectivity in mammals and birds: a hypothesis. Brain Res Bull. 2006;69(1):20–9.

Rial RV, Akaârir M, Gamundí A, Nicolau C, Garau C, Aparicio S, Tejada S, Gené L, González J, De Vera LM, Coenen AM, Barceló P, Esteban S. Evolution of wakefulness, sleep and hibernation: from reptiles to mammals. Neurosci Biobehav Rev. 2010;34(8):1144–60.

Riedner BA, Vyazovskiy VV, Huber R, Massimini M, Esser S, Murphy M, Tononi G. Sleep homeostasis and cortical synchronization: III. A high-density EEG study of sleep slow waves in humans. Sleep. 2007;30(12):1643–57.

Riedner BA, Hulse BK, Murphy MJ, Ferrarelli F, Tononi G. Temporal dynamics of cortical sources underlying spontaneous and peripherally evoked slow waves. Prog Brain Res. 2011;193: 201–18.

Ringli M, Huber R. Developmental aspects of sleep slow waves: linking sleep, brain maturation and behavior.Prog Brain Res. 2011;193:63–82.

Robinson TE. Hippocampal rhythmic slow activity (RSA; theta): a critical analysis of selected studies and discussion of possible species-differences. Brain Res. 1980;203(1):69–101.

Ruskin DN, Bergstrom DA, Kaneoke Y, Patel BN, Twery MJ, Walters JR. Multisecond oscillations in firing rate in the basal ganglia: robust modulation by dopamine receptor activation and anesthesia. J Neurophysiol. 1999;81(5):2046–55.

Schabus M, Dang-Vu TT, Albouy G, Balteau E, Boly M, Carrier J, Darsaud A, Degueldre C, Desseilles M, Gais S, Phillips C, Rauchs G, Schnakers C, Sterpenich V, Vandewalle G, Luxen A, Maquet P. Hemodynamic cerebral correlates of sleep spindles during human non-rapid eye movement sleep. Proc Natl Acad Sci USA. 2007;104(32):13164–9.

Schoenenberger GA, Maier PF, Tobler JH, Monnier M. A naturally occurring delta-EEG enhancing nonapeptide in rabbits. X. Final isolation, characterization and activity test. Pflugers Arch. 1977;369(2):99–109.

Steriade M. Grouping of brain rhythms in corticothalamic systems. Neuroscience. 2006; 137(4):1087–106.

Steriade M, Contreras D, Curró Dossi R, Nuñez A. The slow (<1 Hz) oscillation in reticular thalamic and thalamocortical neurons: scenario of sleep rhythm generation in interacting thalamic and neocortical networks. J Neurosci. 1993a;13(8):3284–99.

Steriade M, Nuñez A, Amzica F. Intracellular analysis of relations between the slow (<1 Hz) neocortical oscillation and other sleep rhythms of the electroencephalogram. J Neurosci. 1993b;13(8):3266–83.

Steriade M, Nuñez A, Amzica F. A novel slow (<1 Hz) oscillation of neocortical neurons in vivo: depolarizing and hyperpolarizing components. J Neurosci. 1993c;13(8):3252–65.

Stern JM, Caporro M, Haneef Z, Yeh HJ, Buttinelli C, Lenartowicz A, Mumford JA, Parvizi J, Poldrack R. A functional imaging of sleep vertex sharp transients. Clin Neurophysiol. 2011;122(7):1382–6.

Tinguely G, Finelli LA, Landolt HP, Borbély AA, Achermann P. Functional EEG topography in sleep and waking: state-dependent and state-independent features. Neuroimage. 2006;32(1): 283–92.

Tobler I, Borbély AA. The effect of 3-h and 6-h sleep deprivation on sleep and EEG spectra of the rat. Behav Brain Res. 1990;36(1–2):73–8.

Trachsel L, Dijk DJ, Brunner DP, Klene C, Borbély AA. Effect of zopiclone and midazolam on sleep and EEG spectra in a phase-advanced sleep schedule. Neuropsychopharmacology. 1990;3(1):11–8.

Vanderwolf CH. Hippocampal electrical activity and voluntary movement in the rat. Electroencephalogr Clin Neurophysiol. 1969;26(4):407–18.

Vanhatalo S, Tallgren P, Andersson S, Sainio K, Voipio J, Kaila K. DC-EEG discloses prominent, very slow activity patterns during sleep in preterm infants. Clin Neurophysiol. 2002;113(11): 1822–5.

Vanhatalo S, Palva JM, Holmes MD, Miller JW, Voipio J, Kaila K. Infraslow oscillations modulate excitability and interictal epileptic activity in the human cortex during sleep. Proc Natl Acad Sci USA. 2004;101(14):5053–7.

Walter WG. The location of cerebral tumours by electro-encephalography. Lancet. 1936;228 (5893):305–8.

Webb WB, Agnew Jr HW. Stage 4 sleep: influence of time course variables. Science. 1971; 174(4016):1354–6.

Chapter 8
The Need of Slow Wave Activity and Cognitive Functions

Abstract According to the synaptic homeostasis hypothesis of Tononi and Cirelli (Sleep Med Rev 10(1):49–62, 2006), the homeostatic regulation of sleep slow wave activity is related to the amount of the synaptic potentiation that has accumulated during the preceding waking state. The homeostatic increase of slow wave activity is shown to be valid for regional involvement in special localization related tasks, especially true for the frontal lobe.

High synaptic potentiation characterizes the early childhood's abundant plastic changes when sleep contains high amounts of delta activity, while the decrease of potentiation in the elderly is associated with important decrease of sleep slow waves, nicely supporting the hypothesis.

According to this hypothesis, slow waves promote a generalized depression or downscaling of synaptic strength reached during wakefulness. Very much in congruence with the hypothesis, it was found that the cerebral blood flow is low in the morning after a night sleep compared to the end of a waking day, as measured by $H_2{}^{15}O$ PET studies. Furthermore, the blood flow values proved to be less and less parallel with the decrease of slow wave activity along the sleep cycles.

Sleep deprivation results in well-known negative cognitive symptoms. That also fits into the hypothesis because of the synaptic overload without the possibilities of downscaling because of the lack of sleep.

A close relationship between the amount of sleep slow wave activity and the cognitive performances in different human pathological conditions also supports the hypothesis.

Sleep deprivation mimics the prefrontal symptoms of mental deterioration in the elderly, where the rebound in frontal delta activity after sleep deprivation is missing. Another aspect of the relationship between human cognitive functions and sleep slow wave activity is that disorders like sleep apnea, Alzheimer disease, or insomnia, all associated with different degree of cognitive decline, show impairment of both NREM sleep and frontal slow wave activity.

The evidences about the importance of NREM slow wave activity in cognitive functions explains the interest about the role of CAP A_1 phenomenon in cognition. We present here studies pointing to relationship between cognitive employment and

P. Halász, R. Bódizs, *Dynamic Structure of NREM Sleep*,
DOI 10.1007/978-1-4471-4333-8_8,

CAP A_1 type amount during the next night sleep. These preliminary results again connect input-dependent slow wave regulation with the use of dependent long-term homeostatic regulation.

Keywords Synaptic homeostasis hypothesis • Slow waves • Delta activity Cognitive functions • Sleep deprivation • Synaptic potentiation

The wisdom of "sleep on it" is confirmed.

8.1 The Need of Slow Wave Sleep

Slow waves were the protagonists of the previous chapters. Why and for which function are they so important? The most coherent and well-established hypothesis of sleep function is presently the so-called synaptic homeostasis hypothesis (Tononi and Cirelli 2006), which places just the slow wave homeostasis in the center of the concept. The hypothesis starts with the well-evidenced assumption that due to the production of a large amount of LTP owing to waking activity, synaptic potentiation increases in many cortical circuits. A necessary requisite for LTP production is the presence of a neuromodulatory milieu, where the firing of presynaptic neurons is followed by the depolarization and firing of postsynaptic neurons. This condition is fulfilled during wakefulness when continuous streaming of impulses impinges on cortical neurons innervated by the ascending arousal systems missing during NREM sleep.

The next statement of the hypothesis is that the homeostatic regulation of sleep slow wave activity is tied to the amount of the synaptic potentiation of the preceding waking state. The higher the amount of potentiation in cortical circuits during wakefulness, the higher is the increase of slow wave activity during subsequent sleep. Several studies have proved that not only the duration of wakefulness but the induction of synaptic potentiation by use-dependent tasks – as it was shown previously – that is responsible for the homeostatic drive increasing slow wave activity (Kattler et al. 1994; Huber et al. 2004; Vyazovskiy et al. 2000). The homeostatic increase of slow activity is shown to be valid for regional involvement in special localization related tasks, and especially the frontal lobe proved to be sensitive for this homeostatic drive (Cajochen et al. 1999; Finelli et al. 2001). The association of synaptic potentiation with developmental periods of early childhood when plastic changes are the most abundant, and high amount of sleep delta activity while decay of potentiation in the elderly associated with important decrease of sleep deltas, fits very well into the hypothesis.

Wakefulness-related molecular mechanisms of a net increase in synaptic weights by LTP are related to enhanced noradrenaline release, enhanced brain-derived neurotrophic factor (BDNF) release, and increased the glutamine receptor1 (GluR1) containing α-amino-3-hydroxy-5-methyl-4-isoxazolepropionic acid (AMPA) receptors in the synaptoneurosomes, as well as to some specific patterns of phosphorylation of GluR1 and calcium-/calmodulin-dependent kinase II (CamKII). Noradrenaline was shown to enhance LTP and to be preferentially released during wakefulness. Accordingly, noradrenaline-depleted rats show a significantly dampened increase of the sleep slow oscillation after prior sleep deprivation (Cirelli et al. 2005). Concerning BDNF, it was repeatedly shown that BDNF is involved in LTP and that the levels of BDNF are higher during wakefulness than during sleep (Vyazovskiy et al. 2008). Moreover, BDNF release is increased in association with exploratory behavior and strongly correlates with the sleep deprivation-induced increase in subsequent sleep slow wave activity (Huber et al. 2007). BDNF was shown to have a causal role in sleep homeostasis and to be involved in local sleep regulation, as local microinjections induced hemisphere-specific increases in sleep slow wave activity. Effects were specific to NREM sleep and did not occur in REM sleep or wakefulness (Faraguna et al. 2008). Functional polymorphisms of the gene encoding the BDNF receptor tyrosine kinase B affect the time spent in slow wave sleep, slow wave EEG activity during NREM sleep (Bachmann et al. 2012), as well as the performance in some neuropsychological tests also impaired by sleep deprivation (Egan et al. 2003; Pezawas et al. 2004). Last, but not least, the changes in the GluR1 containing AMPA receptors and CamKII are the core elements of LTP induction in vivo. Their changes were shown to parallel the changes in the homeostatic sleep need assessed by electrophysiological and behavioral methods (Vyazovskiy et al. 2008). A mechanism of use-dependent GluR1 AMPA receptor increase was proposed by Krueger et al. (2008). Authors are arguing that presynaptic neuronal firing is associated with ATP-release, which binds in part to the P2R purine receptors on the glial cells. In turn glial cells release somnogenic cytokines, like TNFα and IL1β. These cytokines activate postsynaptic enzymatic activities in relation with the NF-κB enzyme. The outcome of this activation is the increase in GluR1 AMPA receptors and adenosine A_1 receptors on the postsynaptic membrane. This cascade of events provides a plausible explanation for the use-dependent sleep regulation process. Although Krueger et al. (2008) argue that sleep homeostasis per se is an emergent property of a large set of neuronal groups (columns), the question of a central sleep-inducing system is still an open one. (We refer to the earlier presented results of the McCarley group showing that local increases in adenosine near the basal forebrain arousal centers can lead to global sleep processes. A similar argument for the possible existence of a global sleep-inducing center and sleep homeostasis is related to the specific involvement of the VLPO region in inducing sleep as well as to the experimental evidence suggesting that local adenosine injection to the VLPO region can induce global sleep.)

The next question is about the function of the sleep slow wave activity. According to the synaptic homeostasis hypothesis, slow waves promote a generalized depression or downscaling of synaptic strength reached during wakefulness.

The exponential decrease of slow wave activity across the sleep cycles described by the Borbély group represents strong electrophysiological fingerprint of this downscaling process. Downscaling probably represents a certain cleaning and refreshing synaptic capacities for new learning. Extended wakefulness or sleep deprivation does not only lead to a net increase in synaptic weights but also to a concomitant occlusion of the LTP process. This is probably due to the saturation of the neural networks (Vyazovskiy et al. 2008). The partial loss of consciousness during sleep is advantageous excluding any interference with the downscaling process.

Very much in congruence with the hypothesis, the cerebral blood flow measured by $H_2{}^{15}O$ PET studies showed a decrease in the flow in the morning after a night sleep compared with the values of the end of a waking day (Braun et al. 1997). Furthermore, the blood flow values proved to be less and less parallel with the decrease of slow wave activity along the sleep cycles.

Sleep deprivation results in the well-known negative cognitive symptoms which also fit into the expectation of the hypothesis because of the synaptic overload without downscaling possibilities. Insomnia associated with hyperarousal also impairs synaptic homeostasis and, consequently, results in fatigue, concentration difficulties, cognitive impairment, and irritability. Depression which is epidemiologically related to insomnia and associated with memory disturbances is also related to the so-called hypofrontality, witnessed by neuroimaging studies, and disrupted sleep pattern (Mobascher et al. 2009). Sleep deprivation which accumulates slow wave activity brings transitory benefit for these symptoms. The ever enigmatic effect of electroshock therapy may get an explanation by the synaptic homeostasis hypothesis as well since the therapeutic effect might be due to artificially induced synaptic up- and downscaling.

The neurophysiologic mechanism of synaptic downscaling and its relationship with slow waves is poorly understood. It is evident that long-term depression (LTD) is an important process taking place during sleep and taking part in the synaptic homeostasis by downscaling. Molecular signs of LTD were revealed to be higher during sleep than during wakefulness, which is the reverse of what we see during wakefulness. The glutamatergic AMPA receptors containing the GluR1 subunit were shown to be removed from the synaptoneurosomes during LTD. Indeed, these receptors are downregulated during sleep compared to periods of wakefulness. A specific pattern of phosphorylation of the receptor suggests that it is not only the low LTP, but indeed the presence of LTD during sleep that is involved in these differences. Thus, LTD seems to be the mechanism of synaptic downscaling taking place during sleep (Vyazovskiy et al. 2008).

Increasing synaptic weights during wakefulness impinge on neurons to fire in synchrony. This results in increased amplitudes and steeper slopes of the slow waves at the beginning of sleep. Experimental evidence shows that induction of repetitive burst pairings in layer V pyramidal cells of the rat is followed by LTD, suggesting a mechanism by which synaptic inputs are proportionally down sized during periods of slow wave sleep (Czarnecki et al. 2007). The process is self-limiting, as LTD reduces synaptic strengths, thus reducing synchronous firing and synchronous slow waves.

8.2 Physiological and Pathological Human Evidences for the Relationship Between Frontal Slow Waves and Frontal Cognitive Functions

There are accumulated human physiological and clinical evidences of interrelationship between slow wave activity of sleep and evolution versus decay of cognitive functions during the human life span. In childhood, delta activity has a higher proportion in sleep and it decreases in parallel with the end of developmental milestones (Darchia et al. 2007). We summarized in the Sect. 7.6 how the delta and/or slow EEG activity during sleep is related to the ontogeny of neuronal systems during childhood and adolescence. Within slow wave activity, the reactive delta portion in the form of CAP A phases follows also significantly the age development as it was described in Sect. 3.6.

In the senium sleep duration (Klerman and Dijk 2008), the amount of sleep delta decreases, especially during the first cycles (Landolt et al. 1996; Landolt and Borbély 2001). Sleep deprivation mimics the prefrontal symptoms of mental deterioration in the elderly (Harrison et al. 2000), and there is diminished frontal delta rebound after sleep deprivation in this age group (Münch et al. 2004).

Chronic insomnia has a negative effect on working memory (Hauri 1997; Boufidis et al. 2004), especially in the elderly (Haimov et al. 2008). Obstructive sleep apnea is harmful for frontal cognitive functions (Mathieu et al. 2008), and SPECT studies in this disorder have shown frontal hypoperfusion (Köves et al. 2003; Puertas et al. 2004). In dementia due to alcoholism and Alzheimer disease, the amplitude of K-complexes and their elicitibility decreases (Colrain et al. 2009).

8.3 CAP and Cognitive Functions

CAP has been introduced as a buffer mechanism counterbalancing the perturbation of NREM sleep, reflecting a sleep-promoting effect within the frameworks of the reciprocal inhibitory dynamics of sleep- and wake-promoting structures. At the same time, however, CAP A phase and especially A_1 considerably contributes to sleep delta activity. We have learned that sleep delta activity is under use-dependent homeostatic regulation. A further development is that the homeostatic increase of slow waves is related to improvement of cognitive functions (Tononi and Cirelli 2006; Huber et al. 2008; Massimini et al. 2009). In this context, the reactive slow wave activity within the frame of CAP sequences became especially interesting. We hypothesized previously that CAP reactive slow wave elements represent an instant homeostatic regulation by which the slow wave equilibrium is balanced during the same night sleep. The existence of something like "slow wave need" during a night sleep is supported by the study of Dijk et al. (1987). He deprived sleep during the first 3 h of the night in young healthy subjects and observed an increase in power densities in the frequency range from 1 to 7 Hz during the subsequent hours of sleep, compared to the same period of the baseline night. In the situation when the rate of CAP increases during a disturbed night sleep, the disturbing intervention does not lead to sleep loss; it only "threats" the evolution of slow wave sleep. It is an interesting question, whether this instant homeostatic regulation also results in cognitive improvement of the next day, as seen in experiments where the increase of delta sleep has that effect.

Whether the homeostatic role of CAP in cognitive functioning has detectable traces immediately during the same night sleep of the disturbance is unanswered. However, Ferri et al. (2008) have found that when persons were exposed to cognitive tasks, the number of CAP A_1 subtypes/h of NREM sleep on the following night was increased, and this increase correlated positively with post-sleep performance. In another study on young human subjects, the same working group (Aricò et al. 2010) published that CAP A_1 subtypes correlated positively with the next-day performance in different neuropsychological tasks, while CAP A_2 and A_3 subtypes correlated negatively with performance. These results mean that the reactive delta compartment of CAP took part in the specific (cognitive) task-related homeostatic regulation and had a simultaneous role in sleep-related cognitive processes.

CAP A_1 phase seems to be an obvious link between responsively of sleep and sleep homeostasis driven by metabolic needs of the brain-fuelling cognitive functions. CAP A_1 phase is a tool by which the sleeper "using up" external influences in the building up of slow wave sleep paves the way for synaptic decay and plastic changes.

References

Aricò D, Drago V, Foster PS, Heilman KM, Williamson J, Ferri R. Effects of NREM sleep instability on cognitive processing. Sleep Med. 2010;11(8):791–8.

Bachmann V, Klaus F, Bodenmann S, Schäfer N, Brugger P, Huber S, Berger W, Landolt HP. Functional ADA polymorphism increases sleep depth and reduces vigilant attention in humans. Cereb Cortex. 2012;22(4):962–70.

Boufidis S, Rikou K, Karlovassiou A, Vlahoyanni E, Balyannis SI, Kosmoidis MH. Impact of insomnia on working memory. J Sleep Res. 2004;13(Suppl s1):91.

Braun AR, Balkin TJ, Wesenten NJ, Carson RE, Varga M, Baldwin P, Selbie S, Belenky G, Herscovitch P. Regional cerebral blood flow throughout the sleep-wake cycle. An H2(15)O PET study. Brain. 1997;120(Pt 7):1173–97.

Cajochen C, Foy R, Dijk DJ. Frontal predominance of a relative increase in sleep delta and theta EEG activity after sleep loss in humans. Sleep Res Online. 1999;2(3):65–9.

Cirelli C, Huber R, Gopalakrishnan A, Southard TL, Tononi G. Locus ceruleus control of slow-wave homeostasis. J Neurosci. 2005;25(18):4503–11.

Colrain IM, Crowley KE, Nicholas CL, Padilla M, Baker FC. The impact of alcoholism on sleep evoked delta frequency responses. Biol Psychiatry. 2009;66(2):177–84.

Czarnecki A, Birtoli B, Ulrich D. Cellular mechanisms of burst firing-mediated long-term depression in rat neocortical pyramidal cells. J Physiol. 2007;578(Pt 2):471–9.

Darchia N, Campbell IG, Tan X, Feinberg I. Kinetics of NREM delta EEG power density across NREM periods depend on age and on delta-band designation. Sleep. 2007;30:71–9.

Dijk DJ, Beersma DG, Daan S, Bloem GM, Van den Hoofdakker RH. Quantitative analysis of the effects of slow wave sleep deprivation during the first 3 h of sleep on subsequent EEG power density. Eur Arch Psychiatry Neurol Sci. 1987;236(6):323–8.

Egan MF, Kojima M, Callicott JH, Goldberg TE, Kolachana BS, Bertolino A, Zaitsev E, Gold B, Goldman D, Dean M, Lu B, Weinberger DR. The BDNF val66met polymorphism affects activity-dependent secretion of BDNF and human memory and hippocampal function. Cell. 2003; 112(2):257–69.

Faraguna U, Vyazovskiy VV, Nelson AB, Tononi G, Cirelli C. A causal role for brain-derived neurotrophic factor in the homeostatic regulation of sleep. J Neurosci. 2008;28(15):4088–95.

Ferri R, Huber R, Aricò D, Drago V, Rundo F, Ghilardi MF, Massimini M, Tononi G. The slow-wave components of the cyclic alternating pattern (CAP) have a role in sleep-related learning processes. Neurosci Lett. 2008;432(3):228–31.

Finelli LA, Borbély AA, Achermann P. Functional topography of the human nonREM sleep electroencephalogram. Eur J Neurosci. 2001;13(12):2282–90.

Haimov I, Hanuka E, Horowitz Y. Chronic insomnia and cognitive functioning among older adults. Behav Sleep Med. 2008;6(1):32–54.

Harrison Y, Horne JA, Rothwell A. Prefrontal neuropsychological effects of sleep deprivation in young adults–a model for healthy aging? Sleep. 2000;23(8):1067–73.

Hauri PJ. Cognitive deficits in insomnia patients. Acta Neurol Belg. 1997;97(2):113–7.

Huber R, Ghilardi MF, Massimini M, Tononi G. Local sleep and learning. Nature. 2004; 430(6995):78–81.

Huber R, Tononi G, Cirelli C. Exploratory behavior, cortical BDNF expression, and sleep homeostasis. Sleep. 2007;30(2):129–39.

Huber R, Määttä S, Esser SK, Sarasso S, Ferrarelli F, Watson A, Ferreri F, Peterson MJ, Tononi G. Measures of cortical plasticity after transcranial paired associative stimulation predict changes in electroencephalogram slow-wave activity during subsequent sleep. J Neurosci. 2008;28(31): 7911–8.

Kattler H, Dijk DJ, Borbély AA. Effect of unilateral somatosensory stimulation prior to sleep on the sleep EEG in humans. J Sleep Res. 1994;3(3):159–64.

Klerman EB, Dijk DJ. Age-related reduction in the maximal capacity for sleep–implications for insomnia. Curr Biol. 2008;18(15):1118–23.

Köves P, Szakács Z, Bernáth I. T99m HMPAO rCBF SPECT findings in OSAS patients. Sleep. 2003;26(Suppl):A232.

Krueger JM, Rector DM, Roy S, Van Dongen HP, Belenky G, Panksepp J. Sleep as a fundamental property of neuronal assemblies. Nat Rev Neurosci. 2008;9(12):910–9.

Landolt HP, Borbély AA. Age-dependent changes in sleep EEG topography. Clin Neurophysiol. 2001;112(2):369–77.

Landolt HP, Dijk DJ, Achermann P, Borbély AA. Effect of age on the sleep EEG: slow-wave activity and spindle frequency activity in young and middle-aged men. Brain Res. 1996;738(2): 205–12.

Massimini M, Tononi G, Huber R. Slow waves, synaptic plasticity and information processing: insights from transcranial magnetic stimulation and high-density EEG experiments. Eur J Neurosci. 2009;29(9):1761–70.

Mathieu A, Mazza S, Décary A, Massicotte-Marquez J, Petit D, Gosselin N, Malo J, Montplaisir J. Effects of obstructive sleep apnea on cognitive function: a comparison between younger and older OSAS patients. Sleep Med. 2008;9(2):112–20.

Mobascher A, Arends M, Eschweiler GW, Brinkmeyer J, Agelink MW, Kornischka J, Winterer G, Cordes J. Biological correlates of prefrontal activating and temporoparietal inhibiting treatment with repetitive transcranial magnetic stimulation (rTMS). Fortschr Neurol Psychiatr. 2009;77(8):432–43.

Münch M, Knoblauch V, Blatter K, Schröder C, Schnitzler C, Kräuchi K, Wirz-Justice A, Cajochen C. The frontal predominance in human EEG delta activity after sleep loss decreases with age. Eur J Neurosci. 2004;20(5):1402–10.

Pezawas L, Verchinski BA, Mattay VS, Callicott JH, Kolachana BS, Straub RE, Egan MF, Meyer-Lindenberg A, Weinberger DR. The brain-derived neurotrophic factor val66met polymorphism and variation in human cortical morphology. J Neurosci. 2004;24(45):10099–102.

Puertas FJ, Tembl A, Cordero JM, Azzi H, Romero MF, Merino MA. Changes in cerebral perfusion studied by PECT in patients with severe obstructive apnea/hypopnea syndrome. J Sleep Res. 2004;13(Suppl s1):592.

Tononi G, Cirelli C. Sleep function and synaptic homeostasis. Sleep Med Rev. 2006;10(1):49–62.

Vyazovskiy V, Borbély AA, Tobler I. Unilateral vibrissae stimulation during waking induces interhemispheric EEG asymmetry during subsequent sleep in the rat. J Sleep Res. 2000;9(4): 367–71.

Vyazovskiy VV, Cirelli C, Pfister-Genskow M, Faraguna U, Tononi G. Molecular and electrophysiological evidence for net synaptic potentiation in wake and depression in sleep. Nat Neurosci. 2008;11(2):200–8.

Chapter 9
Overview

9.1 Summary

In this book, we provide a new approach to sleep regulation, incorporating the new knowledge of use-dependent homeostatic regulation associated with plastic changes during NREM sleep. We amalgamated this homeostatic regulation with the line of the parallel developed by microstructural dynamics driven by sensory impulses from the surrounding world. One of the main lines in the book is the recognition of input-dependent (reactive) slow wave activity (responses) incorporated as "instant" homeostatic regulation in the slow wave homeostatic process. Consequently, we have shown the state dependency of sleep or arousal promoting effect of sensory input during NREM sleep, and the interdigitation of micro- and macro-level of sleep dynamism. This may explain how the sleeper is kept in contact with the surrounding world, preserving the continuity of sleep.

Our scenario for the dynamic NREM sleep regulation can be summarized as follows: The homeostatic regulation is fed by the synaptic potentiation during the day providing a demand for producing slow waves, allowing synaptic decay. This is the main force of NREM sleep. When the homeostatic pressure is high, it drives the hypothalamic sleep-promoting neurons to high level of firing. The high firing rate of these neuronal assemblies keeps the wake-promoting arousal systems under inhibition. The inhibition of the arousal system liberates the thalamic burst-firing system which produces spindling and slow wave oscillations (Fig. 9.1). In this condition, sensory input fuels phasic slow wave activity reflected by the CAP A_1 phases, providing slow wave injections to each perturbation of sleep (instant homeostatic regulation) during the "D" slopes characterized by high homeostatic pressure. During the decrease of homeostatic pressure (synaptic decay), the firing of the VLPO falls gradually down the impinging sensory input from the external world, reflected by CAP A_{2-3} phasic events, it starts to bring the sleeper to higher and higher level of arousal along the "A" slope preparing the next REM sleep (Fig. 9.2).

The cooperation of metabolic demand-driven long-term homeostatic regulation and the input-driven short-term phasic arousal-related regulation provide flexible

P. Halász, R. Bódizs, *Dynamic Structure of NREM Sleep*,
DOI 10.1007/978-1-4471-4333-8_9, © Springer-Verlag London 2013

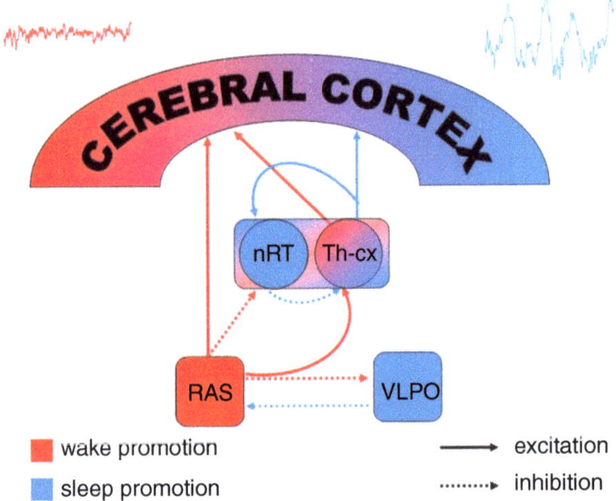

Fig. 9.1 Mutual inhibition between the sleep (*blue*) and arousal (*red*) systems as well as their effect on the thalamocortical circuitry. (I) Awake time synaptic potentiation elevates homeostatic pressure, elevating firing rate in VLPO and other sleep promoting neurons. (II) VLPO and other sleep promoting neurons inhibit the arousal system (*RAS*). (III) Inhibition of the arousal system liberates the thalamic burst-firing system producing spindle and slow wave oscillation (top trace in blue) instead of low amplitude high frequency EEG (top trace in red). *VLPO* ventro-lateral preoptic area, *RAS* reticular activating system, *Th-cx* thalamocortical neurons, *nRT* nucleus reticularis thalami, *continuous* and *broken arrows* symbolize excitation and inhibition, respectively, *red* and *blue* colors represent wake and sleep promotion, respectively

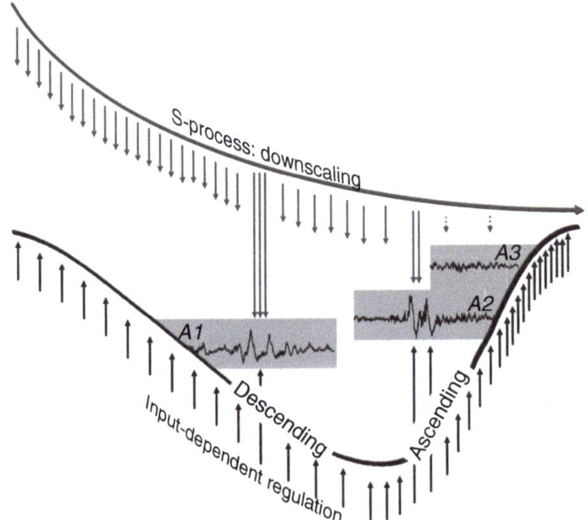

Fig. 9.2 Input-dependent phasic sleep regulation during the descending and ascending slopes of sleep cycles under the influence of homeostatic regulation. During the descending slope while homeostatic pressure is high phasic sensory input (*black arrows*) evokes A_1 type sleep-like responses providing delta injections (*gray arrows*) and maintaining of sleep. Contrasting during the ascending slope A_2 and A_3 type responses promotes the arousal process and brings the sleeper nearer to the next REM phase that is no more counterbalanced by the decreasing homeostatic pressure

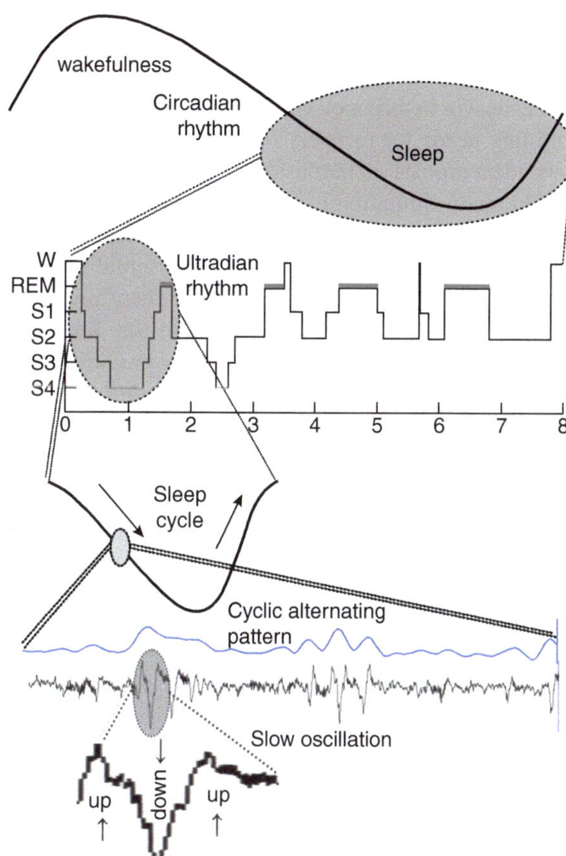

Fig. 9.3 Yin-yang-like representation of the antagonistic twin principle in connection to and separation from the environment on different levels of sleep/wake continuum. *Upper row*: within the circadian rhythm waking states alternates with sleep periods. *Second row*: within the ultradian rhythm in each sleep cycle the descending slope represents sleep-promoting conditions and the ascending slope wake-/REM-promoting conditions. *Third row*: cyclic alternating pattern represents microstructural activation/inactivation episodes with input-related sleep-promoting phasic events. *Fourth row*: <1 Hz slow oscillation up-and-down states represent within some hundred milliseconds alternating almost wake state-like neural/synaptic activation with severe disfacilitation without neural/synaptic activity

adaptation to accomplish separation from the external world to fulfill trophotropic and keep the contact with it for accomplishing ergotrophic activity.

The same general principle of sleep/wake regulation seems to be represented by the yin-yang alternation of high- and low-level activity where the high-level activity connects the brain with the external world, while the low-level activity provides separation.

This alternation in the link with the environment can be followed on different levels (Fig. 9.3). First, we have sleep and wakefulness as a covering curve of these alternations. Within sleep, we have the sleep cycles with alternation of the "D" and "A" slopes of sleep cycles and the arousal- and sleep-like phasic events in coalescence as smaller grade fluctuations driven by the input factors. Lastly, within slow wave oscillation the up-and-down states represent the same alternation of connection and separation.

The self-identical repetition of the alternation of connection and separation on larger and smaller levels show certain resemblance to the dynamics of fractal processes.

9.2 Endeavors, Achievements, and Limitations

Our endeavor in this book was to connect the metabolic driven homeostatic regulation, fulfilling needs for recovery of the cortex for cognitive functioning with the input-dependent arousal-/antiarousal-related regulation making sleep an open system.

The main limitation to build up a thoroughly experimentally proven regulation system is the gap between the EEG/behavior level and the level of neuronal firing patterns in the hypothalamic neuronal assemblies, constituting the reciprocal antagonistic sleep/wake neuronal network. Therefore, the link between the two levels is arbitrary and hypothetical. Although the Takahashi group from Japan and the Harvard and Lyon sleep research groups provide certain data about an existing relationship between sleep/wake EEG and brain stem neuronal firing patterns, however, this relationship remains still hypothetical. Another possibility is that the phasic EEG events are not underlain by brain stem neuronal changes being responsible only for more tonic modifications of the sleep/wake balance during the sleep cycle, and phasic events are purely cortical. To prove or reject the hypothetical interrelationship between EEG phasic events and brain stem neuronal firing, further experimental studies are warranted.

9.3 Outlook

It seems to be obvious that sleep is a price of something what we need. Probably, the prices of plasticity as Tononi and Cirelli have proposed. Looking back to the phylogenesis of sleep, we have seen that the biggest achievement by the jump from the reptile existence to the mammals was the development of the cholinergic arousal system, providing more complete wakefulness. With the discovery of >1 Hz slow oscillation, it was revealed that during the negative and positive half waves of slow oscillation, an alternating sleep- and wake-like connection with the environment is possible, nearly simultaneously in the mammalian brain. This development could be considered as something in preparation. It might be a start to conquer the brain from sleep to a more wakeful life just in the deepest part of sleep. Men spend one-third of their life in sleep as a must for endophylactic tasks provided by it. Potentially, the up-and-down states through a future development shall be able to ensure parallel local waking states, weaving further the tissue of our life within the frame of sleep, while other parts deal with recovery functions.

Index

P. Halász, R. Bódizs, *Dynamic Structure of NREM Sleep*,
DOI 10.1007/978-1-4471-4333-8, © Springer-Verlag London 2013

If you have any concerns about our products,
you can contact us on
ProductSafety@springernature.com

In case Publisher is established outside the EU,
the EU authorized representative is:
Springer Nature Customer Service Center GmbH
Europaplatz 3, 69115 Heidelberg, Germany

Printed by Libri Plureos GmbH
in Hamburg, Germany